Understanding the Market Economy

Understanding the Market Economy

ARNE JON ISACHSEN, CARL B. HAMILTON,
and THORVALDUR GYLFASON

OXFORD UNIVERSITY PRESS

1992

Oxford University Press, Walton Street, Oxford OX2 6DP
Oxford New York Toronto
Delhi Bombay Calcutta Madras Karachi
Petaling Jaya Singapore Hong Kong Tokyo
Nairobi Dar es Salaam Cape Town
Melbourne Auckland
and associated companies in
Berlin Ibadan

Oxford is a trade mark of Oxford University Press

Published in the United States
by Oxford University Press, New York

British Library Cataloguing in Publication Data
Data available

Library of Congress Cataloging in Publication Data
Isachsen, Arne J.
Understanding the market economy / Arne J. Isachsen, Thorvaldur
Gylfason, and Carl B. Hamilton.
p. cm.
Includes bibliographical references and index.
1. Economics. 2. Comparative economics. 3. Capitalism.
I. Porvaldur Gylfason, 1951– . II. Hamilton, Carl. III. Title.
HB171.5.I78 1992 330.12'2—dc20 92-24790
ISBN 0-19-877356-0
ISBN 0-19-877357-9 (Pbk)

Set by Hope Services (Abingdon) Ltd
Printed in Great Britain by
Biddles Ltd., Guildford and King's Lynn

To
Jannike and Elisabeth
Mette, Christoffer, and Tobias
Jóhanna and Bjarni

Preface

THE inspiration to write this book came to us during the conference 'The Baltic Family' held in Kaunas, the old capital of Lithuania, in October 1990. Realizing the enormous challenges facing Lithuania in its transition from plan to market, we decided to try our hands at a book explaining the workings of a market economy. A modest contribution to such an understanding is the purpose of this book.

In the course of writing this book, the Centre for Baltic Business Development has been established at the Norwegian School of Management. The present book is the first contribution in the Learning Library Programme of that centre.

Originally, the book was intended for beginners' courses in economics and business administration. However, since we rely on simple examples rather than mathematical equations, the book should be accessible to a much wider audience than a standard textbook in economics. Through circulating earlier drafts of the manuscript among Scandinavian readers, we have indeed learned that people from different walks of life have got a good grasp of economics from reading it. This makes us believe that business people, academics in other fields, politicians, public servants, and enlightened laymen in general could benefit from reading the book.

We thankfully acknowledge useful comments on earlier drafts of the manuscript from Olav Bjerke, Ole Bjørn Fausa, Harry Flam, Knut Isachsen, Gerson Komissar, Preben Munthe, Reidar Nilsen, Øystein Noreng, Mats Persson, Geir-Helge Sjøtrø, Jan Arild Snoen, Øystein Thøgersen, Jens Jonathan Wilhelmsen, and Anders Åslund. Tore Rokstad and Nils Haugland have provided able research assistance. In addition, they patiently updated the manuscript based on three authors rewriting the text.

John Christian Langli, Ph. D. student at the Norwegian School of Management, wrote a draft of a chapter on accounting. With minor modifications this appears as Chapter 17 in the present book. We owe him our gratitude.

Finally, we would like to thank the Department of Foreign Affairs in Norway, the Swedish International Development Agency (SIDA), and the Nordic Economic Research Council for financial support.

The Danish physicist Niels Bohr once remarked that physics is not about the real world because the world is too complicated for that; no, he said, physics is about what can be said about the world. We take the same view of our subject: economics is about what can be said about economic life.

Furthermore, to a large extent, economics is concerned with marginal changes. How much will the demand for shoes decline in response to a 10 per cent increase in the price of shoes? What will be the effects of a 5 per cent devaluation of the currency? How will increased taxation of production that causes pollution accompanied by a commensurate decrease in income tax affect the economy?

However, the economic challenges now facing the nations of Central and Eastern Europe are anything but marginal. When considering these challenges, economists therefore have reason to be modest about what they have to say. We do not claim to present unequivocal or simple solutions to the serious problems that we discuss in this book. Our objective is more modest: to present an intelligible account of basic economic mechanisms and relationships and to consider how they can be brought to bear on the transition from plan to market.

Many people consider economics a difficult field. Precisely for that reason, we have endeavoured to keep the text as simple as possible, without, of course, losing rigour. The extent to which we succeed in conveying a basic understanding of economics is for you to decide, after having studied the chapters that follow. We now put our pens to rest and wish you a happy journey.

A.J.I.
C.B.H.
T.G.

Oslo, Stockholm, and Reykjavík
December 1991

Contents

List of Figures

List of Tables

List of Supplements

1
Introduction

Whether you like it or not, history is on our side. We will bury you.

Nikita Khrushchev

IN 1960 Nikita Khrushchev claimed that in the course of the next twenty years the Soviet Union would have caught up with and surpassed the United States in terms of material standard of living. Add on ten years, and you will find an economic system in a state of dissolution: the Soviet system.

In the thirty years that have passed, the system of economic planning has shown itself unable to deliver what its leaders promised. Recognition of this fact has led to the desire to replace that system now. The market economy is generally regarded as the timely solution.

There is scarcely any doubt that the market economy has shown itself superior to the planned economy as far as economic growth is concerned. The material standard of living in Finland today is four times higher than in the Baltic states, despite the fact that at the start of the Second World War national income per head was about the same in all of these countries. In 1940 the economic situation of Hungary was about the same as that of Austria and the standard of living in Czechoslovakia was higher than in Austria. Today, however, the standard of living for the average Austrian is at least twice as high as for the average Hungarian or Czechoslovakian.

A similar story can be told about other continents. In 1960 the material standard of living in South Korea was certainly not significantly higher than in Tanzania; in 1988, however, the average annual purchasing power of income per inhabitant in Tanzania was $US355, compared with $US3,056 in South Korea, i.e. more than eight times as high. Compared with North Korea, which at the end of the Korean War some forty years ago was the richer part of the Korean peninsula, South Korea's income per inhabitant today is at least five times as high.

Although development of the material standard of living is difficult to compare from country to country, the conclusion remains: the market economy has shown itself to be vastly superior to the planned economy when it comes to economic growth. And the revolution in Eastern and Central Europe in 1989–90 could not be easily understood without reference to the stagnating standard of living in these countries.

Amidst the enthusiasm surrounding liberation from the command economy's tight bonds, and the eagerness to start using 'the new system', it might be opportune to focus on the questions that the introduction of a market economy must raise, as well as pointing to the demands and challenges that society, and individual members of society, will face if the new system is to have a chance of functioning reasonably satisfactorily.

To put it in slightly simpler terms, one could say that the two major problems an economic system must solve are to ensure an effective utilization of resources in the short and long term, and a just distribution of the resultant income from production. In any economic system, the state will inevitably play an important part in solving these problems.

1.1. Efficiency

Efficient use of resources means that everyone who wishes to work will have a job. Full employment is an important objective. In the sense that virtually everyone does have a job in planned economies, economic planning has been more successful than the market system in that regard. However, with respect to how efficiently labour is utilized, the market system is clearly superior. This is due among other things to the fact that in a planned economy an unreasonable amount of working time is spent in idleness on the job and in queues outside shops.

Being idle on the job is a way of wasting resources. In a market economy the capital owner receives the *profit*, namely what is left of income after all other costs have been covered; thus, the owner of capital is given a strong incentive to prevent wastage of labour and other resources.

In the longer term, however, full employment is not merely a question of keeping people in work and ensuring that working hours are utilized effectively. At least as important is what they are working at and how. And here the market system has shown itself to be vastly superior to economic planning. This is because the market adapts itself to changes in technology and preferences.

When new methods of production of well-known products are launched, those enterprises that do not adjust to developments will have problems in selling their goods; quite simply, they risk pricing themselves out of the market. In the short term this may lead to the closure of enterprises and higher unemployment. In the longer term, however, it ensures a great degree of flexibility for the system viewed as a whole.

In a market economy enterprises compete in selling their goods and services. Their main concern is to serve the interests of consumers. This encourages the production of new commodities which better satisfy customers' demands. Innovative enterprises that succeed in this will capture a larger part of the market. Enterprises that are unable to improve the quality of the goods they produce will have to reduce their activity. Despite the fact that the East German Trabant was the cheapest car when East and West Germany were united in 1990, such a poor-quality product was no longer in demand when other cars became available in the eastern part of the country. When the production of poor-quality products falls or ceases, resources are released for expanding enterprises and industries.

If, over a period of time, customers' tastes change, enterprises must quickly make note of this and make every effort to restructure production in the appropriate direction. One of the reasons why Japanese car producers experienced such a great increase in their market share in the United States after the quadrupling of oil prices in 1973–4 was that Japanese manufacturers were much better than their American competitors at meeting people's demands for more fuel-efficient cars. The American car manufacturers needed much more time to adapt to more expensive petrol. Luckily for consumers, there were alternatives to the large American cars in the small cars that were imported from Japan (and also from Europe).

In the long term, then, the market has a built-in mechanism which ensures that the structure of production changes in line with technological progress and with changes in preference among customers.

In a planned economy, production is directed by a central planning body. In principle, this body should be continually revising its plans, on the basis of technological innovations in production and changing tastes of consumers. In practice, it does not work that way. First, the manager of the individual enterprise has little incentive to take the risks involved in introducing new technology. There is no owner or group of owners that harvests the profits of increased productivity. Second, the planned economy is notoriously characterized by queues: the prices of most goods and services are far too low in relation to the demand. This makes it difficult for the planning authority to know how the existing resources can be optimally utilized, in other words what goods and services customers are most interested in obtaining more of.

In practice, economic planning has become a system where realization of the plan *as such* is a prime concern, rather than the satisfaction of customers' demands. Whereas a market economy is consumer-oriented, a planned economy is producer-oriented.

When economic planning managed neither to encourage innovation nor, over a period, to adjust itself to changes in preference among customers, the result was stagnation in economic growth.

1.2. The Role of the State

There is no such thing as a 'pure' market economy—an economy where all economic decisions are made in free markets. All market economies are 'mixed' because in any modern society the state—as representative for the interests of the community—has an important role to play. Although the members of a market economy generally enjoy much greater freedom of action than those in a planned economy, the state must ensure that an appropriate institutional framework exists. This involves a suitable legislative code which ensures among other things that the individual is free to establish his own business. Further, the state must ensure law and order, a suitable military defence and welfare system, and agreements of various types with other countries. Bureaucrats in the public sector have the job of implementing political decisions in these areas. The overall effectiveness of the economy is therefore dependent also upon how well the bureaucracy functions.

When the state shapes the institutional framework and the 'rules of the game' for the business sector, it is important that negative side-effects of economic activity are taken into account, not least of which is environmental damage.

Finally, low inflation, i.e. virtual stability in the general level of prices, will improve the functioning of an economy. However, low inflation does not mean that individual prices should be 'low', or that prices, including rent and wages, should remain unchanged in relation to one another over a period of time. The principal responsibility for inflation will always rest with the authorities. They have a monopoly on the issue of banknotes. And in the long term, the supply of money in the economy will be decisive for the general development of prices.

1.3. Economic Justice

In addition to an effective utilization of resources, an economic system must ensure that there is a just distribution of income. One important reason why economic planning was long held in great respect in many Western circles was that distribution of income under an economic planning system was supposed to be more just than under a market economy.

We shall not attempt to define the term 'just distribution of income' here. We will content ourselves with pointing out that distribution of income in the market economies of Western Europe is probably more even than in the planned economies of Central and Eastern Europe, where an extensive system of privileges had been created by and for party bosses. They had special shops, where they could purchase a wide range of goods cheaply and without queuing. In addition, many of them enjoyed perks such as their own 'dachas' and elegant cars that were out of reach of most other people.

To summarize, economic planning has outperformed the market economy only when it comes to providing people with jobs. When it comes to the task of ensuring satisfactory utilization of resources in the short term—and even more so in the long term—the market has proved superior to planning; at the same time, distribution of the resultant income from production has been no less equitable in market economies than in planned economies.

1.4. Overview of the Book

In the chapters that follow we will be returning to these economic problems. An important objective of the book is to explain *why* things had to go as they did: that is, why each system tended to solve the various problems with varying degrees of success. A better understanding of this will provide a good point of departure for the challenges involved in the transition from a planned to a market economy.

In Part I we consider why economics is justified in having the status of a special discipline: i.e. because of the necessity to manage scarce resources efficiently. We shall spend some time explaining three fundamental concepts: *supply*, *demand*, and *market-clearing*. These terms will be employed to show how the total production of a commodity, e.g. shoes, and the price this product will be sold for are determined in a market economy. We shall also show how these concepts can be used to explain the determination of the general price level and *gross national product* (GNP), which is simply the sum of the quantities produced in individual markets. The problems of inflation and unemployment are touched upon, although they will be analysed more fully later. In order to highlight the mechanisms at work in the various systems, we will make the presentation as simple as possible.

The final chapter in Part I examines how resources are utilized under economic planning. It is important to have an understanding of the system one is changing from, in addition to insight into what one is changing to.

Part II is concerned with the new system; here the functioning of a market economy is studied in some detail. A fundamental prerequisite for a well functioning market is that the individual producer and consumer have access to alternative suppliers; put another way, it requires the absence of a monopoly. Since producers have good grounds for seeking a monopoly position, it is important that an appropriate legislative code exists which will prevent this, and that, via free import, competition exists that can prevent businesses from monopolizing the market.

At least as important, in the present situation of the countries of Central and Eastern Europe, is the establishment of a code of legislation which safeguards the right to private ownership: no right of ownership, no capitalists; no capitalists, no capitalism; and no capitalism, no market economy.

In Part III we attempt to explain terms that are commonly used in a market economy but are less well known in a planned economy. The significance of money in facilitating exchange and trade, the importance of a well functioning capital market to the promotion of saving and investment, and the role of international trade as a guide to the right areas for specialization in production will all be discussed. Further, the way a modern labour market works will be taken up and the economic role of government will be considered.

In Part IV problems pertaining to privatization of the means of production are considered. Economic decision-making presupposes adequate financial information. Thus, we present a brief overview of the elements of accounting. The book concludes with an important and difficult chapter on challenges entailed by the transition from a planned economy to a market economy. It is difficult for us to envisage a gradual, long-drawn-out process; if the market is to function well, it will be necessary for most things to fall quickly into place.

Before we start, we must make one thing clear. This is not a book about political economy. We have no intention of giving a broad historical survey of the forces that have created the present crisis. Instead, we are concerned with explaining, in the simplest way possible, economic ideas and mechanisms that are little utilized and generally poorly understood in Central and Eastern Europe. If the unknown in this way becomes less mysterious, perhaps it may also become less threatening.

Since we emphasize economic mechanisms rather than political economy, it is hard to prevent the presentation from occasionally seeming somewhat naïve, especially at the beginning. That is the price we must pay in our attempt to provide a clear insight into the way the economy works.

For those who believe that the market economy is a magic formula, which, if recited a sufficient number of times, will deliver the goods—and our impression is that many people believe just that—this book may have a sobering effect. In the market system, too, a large number of dilemmas and unsolved problems are inherent, even though they may be less numerous and less severe than under economic planning. The degree to which a country succeeds in its transition from a planned to a market economy will depend upon how well these problems are understood, and upon how palatable the population finds the trade-offs that must take place.

PART I

The Economic Problem

> Every individual endeavours to employ his capital so that its produce may be of greatest value. He generally neither intends to promote the public interest, nor knows how much he is promoting it. He intends only his own security, only his own gain. And he is in this led by an invisible hand to promote an end which was no part of his intention. By pursuing his own interest he frequently promotes that of society more effectually than when he really intends to promote it.
>
> Adam Smith

ECONOMICS is the study of the utilization and distribution of scarce resources. Scarcity arises when there is not enough of something. Prior to the industrial revolution of the eighteenth century, there was sufficient clean air and pure water everywhere; people were able to use air and water as much as they liked without there being an insufficient amount for others. In this sense air and water are *free goods*. They could be freely utilized without cost by everyone. Mostly, they still can.

This is not the case with regard to clothes and food, cars and houses. If society wishes to increase production of cars, for example, there will be less resources left for production of other things—houses, for instance. When more of one thing leads to less of another, we are dealing with *economic goods*.

An economic decision involves choosing one thing (e.g. the utilization of resources on the production of ten cars) at the expense of another (e.g. the production of two houses). For society as a whole as well as for the individual inhabitant, it thus becomes important to weigh *alternatives* against one another. This contraposing of alternatives is one of the most important roles of an economic system.

In a market economy the individual producer will decide what is to be manufactured, how much, the quality, and the price. In a planned economy it is a centralized body—the planning authority—that determines this.

When the quantity of a product that people wish to buy deviates from the quantity that is offered for sale, the question arises of mechanisms that can create a balance. In a market economy the individual producer, on the basis of his desire to maximize profits, will himself continually consider and carry out changes in the quantity produced, the quality, and the price. The planned economy, on the contrary, has shown a very poor capacity for such adaptation. The official prices are often virtually unchangeable. Moreover, changes in the planned supplies both between enterprises and from enterprises to consumers have been shown to be very difficult to achieve. Furthermore, the desire for the right quality at the right price is seldom satisfied.

2

Production and Distribution

> The inherent vice of capitalism is the unequal sharing of blessings; the
> inherent virtue of socialism is the equal sharing of misery.
>
> Winston S. Churchill

ECONOMICS, then, is all about the utilization and distribution of scarce
resources. We could say that the economic system in any society is confronted by two main questions: 'What should be produced?' and 'How
should the resulting goods be distributed?'

2.1. Market or Plan?

The decision about what should be produced is usually made in parallel
with decisions concerning who is to be in charge of production and what
sort of technology is to be utilized. By 'technology' we understand the
organization of the enterprise, as well as what sort of machinery, etc., is to
be used. The quality aimed at and the quantity to be produced must also
be determined.

These questions, which all concern production, must be answered
simultaneously. This applies to a market economy as well as to a planned
economy. In a market economy individual enterprises make the decisions;
in a planned economy it is the 'centre' (or planning bureaucracy) that, by
means of directives and rules, seeks to ensure that factories and other production units will do exactly what the central authorities want.

As far as the other main question is concerned, namely the distribution
of the income resulting from production, the idea behind a planned economy is that this can be decided independently of the question of production. 'All people shall contribute according to ability and receive
according to need' was the goal of the early communists. An attempt is
thus made to separate the decision concerning *distribution* of income from
the decisions concerning *production* of it.

In a market economy, on the other hand, the initial distribution of
income (before taxes and transfers) is determined simultaneously with
production. This factor is precisely the strong point of the market system,

as well as its weakness. An attempt is made via the tax and transfer system to compensate for this weakness. In the regulations concerning taxation of personal income and company profits, it is necessary to make political choices between the desire to achieve maximum production and the desire for a just distribution of the income resulting from production.

The strength lies in the encouragement the individual producer receives when he is able to enjoy the fruits of his own initiative and enterprise. If he is skilful and fortunate, both in the choice of technology and in the choice of the goods he produces, he can enjoy a solid profit; if he is unsuccessful in this he has full grounds, on his own initiative, to consider changes in the product range, technology, quality, and price.

The weakness lies in the fact that fortune and misfortune may easily play a dominating part. And the final result, namely the distribution of income that emerges, may be considered extremely unreasonable by many people. Further discrepancies in the distribution of income are due to the fact that different people have different prospects as a point of departure. Just think of those who are born into rich families, and who, via inheritance, start their working lives with a sizeable fortune—or those who have been assigned a greater portion of intelligence than others, and who for this reason have every possibility of also having a richer life, viewed in purely economic terms.

As we pointed out in Chapter 1, distribution of income in market economies is hardly less even than in planned economies. In relative terms, the distribution of the 'pieces of the pie' seems to be roughly the same under both systems according to available statistics. However, because of far more rapid economic growth under the market than under the planned economy, the 'pieces of the pie' are significantly bigger in absolute terms in market economies.

In order to answer the two questions listed above (what should be produced, and how the resulting goods should be distributed), millions of decisions must be made every day. For example, take the individual household. Almost every day it purchases various goods and services. What should be bought? How much? Or consider the individual enterprise. How, for example, should a shoe manufacturer organize the day's production? What types of shoe should be produced and in what quantities? What raw materials would it be best to utilize? Which workers should do what, and how should the machinery available be utilized most effectively? And in what way should the finished products be distributed to customers?

Taking into account that the questions that must be answered require a number of decisions, the organization of the economy in a society must be viewed on the basis of the mechanisms utilized to co-ordinate all of these decisions.

In a market economy *prices* play a crucial part. Through prices, customers receive information concerning how much money they must sacrifice in order to obtain a product. And producers gain knowledge of how much revenue they can expect by producing and selling products. As we shall see later, all necessary information—in a well functioning market—is inherent in prices. Given that producers are well acquainted with the technology, they will only need to know in addition the prices of inputs (labour, capital, and raw materials) and the prices of the products they themselves produce. With this information to hand, it is up to the individual producer to decide what, how, and how much should be produced of each type of product.

SUPPLEMENT 2.1. PRICES AND INCENTIVES

By 'correct incentives' is meant reward systems which encourage individuals, households, enterprises, and public authorities to take decisions that are most appropriate from the point of view of the national economy. It is important to create incentives that lead a decision-maker to act both in his own best interests, and also in the best interests of society. With the proper incentives, for example, there would be a close link between what was profitable for the individual enterprise and for the national economy as a whole.

Prices influence incentives in a fundamental way. They function best when there is competition. Prices and competition lead producers to be cost-conscious and to take into account that resources can be deployed in alternative manners. At the same time, consumers will, within their budgetary constraint, determine the composition of consumption in a trade-off between the prices of goods and the usefulness they have for the individual. Prices give information regarding production costs and willingness to pay in a market economy.

It is often extremely difficult to design proper incentives, especially when the individual's decision has consequences for others. In cases of such 'external effects', unfettered competition in free markets fails to create correct incentives. Cigarette smoking is a case in point because smokers pollute the air inhaled by non-smokers. Free-market prices of cigarettes are generally considered too low because they encourage people to smoke too much, and thus to pollute the air breathed by others. This is an important reason why cigarettes, and tobacco in general, are subject to special taxation in virtually all market economies. The government intervenes to correct a market failure. It is impossible, however, to determine objectively how high these taxes should be. If smokers were to hurt only themselves by smoking, the case for cigarette taxes would be weaker.

But even when external effects are absent, serious incentive problems may arise. If a division of a company produces more than its budget was calculated for, it will be natural for the company's management to place higher demands the following year. If bonuses or other perquisites are linked to such overproduction, it may be appropriate to limit efforts one year in order to maximize future bonuses. It has, for example, been claimed that public-sector undertakings that have exceeded their budgets have been rewarded with larger budgets the next year, whereas savings have led to reduced allocations.

In a planned economy the exceeding of production goals in one period will give a considerable bonus for that period. However, the planned production figures for the following year will probably be raised, and the workers must resign themselves to working harder still. This may hinder increased efforts in the first place.

Incentive mechanisms can have unexpected effects which may often be hard to detect in advance, but may seem rather obvious in retrospect. In order to intensify the hunt for smugglers in the Fujian Province of China, it was thought to be a good idea to let the customs officials keep a certain proportion of the smuggled goods they managed to confiscate. However the system worked in direct contravention of its purpose: when customs officials managed to catch smugglers, they made sure that they were immediately released in the hope of obtaining even more smuggled goods as a bonus later on.

Source: NOU (1988:21, p. 127)

Since the individual producer makes all decisions concerning his own production, a market economy is a *decentralized* decision-making system. In a planned economy, on the other hand, the individual producer is informed by the central authority of what is to be produced and how production is to be distributed—for example as production materials in other enterprises or as supplies to shops, for sale to the consumer. It is not enough for the central authority to give directives concerning production of the finished products: supplies from enterprise to enterprise also are subject to decisions made by the central authority.

Whereas prices are carriers of information to enterprises in a market economy, explicit figures on physical production, given in a plan devised by the central authority, play this part in a planned economy. For this reason a system of economic planning is also called a *centrally planned economy* or a *command economy*.

2.2. *The Wealth of Nations*

A look at the past tells us that it is only in the course of the last couple of centuries that the world has experienced relatively even economic growth over such a long period. It all started with the industrial revolution in the British Isles around 1750. The first 'modern' textbook on economics was written by the Scotsman Adam Smith, who was a professor of moral philosophy. His *Wealth of Nations*, which came out in 1776, still constitutes a rich source of insight into economics.

Adam Smith attributed economic progress first and foremost to the division of labour. By dividing production of (for example) shoes into a number of operations, it was possible to increase the speed of each operation. When some workers specialize in cutting leather while others become expert at hammering on soles, work is performed more rapidly. A shoe manufacturer with fifty employees can produce perhaps 200 times as many pairs of shoes as a shoemaker who works alone. In that case the larger, specialized enterprise will be four times as productive as the small one, where one man must do all the jobs.

Furthermore, division of labour creates the opportunity to employ more specialized machines, which contribute further to increasing the efficiency of the individual worker. However, mechanization means that some citizens must be given freedom to produce new capital equipment. This implies that the production of consumer goods by the remaining members of the community must also meet the consumption needs of those who produce the new machines.

Finally, someone must *own* the machines and the rest of the production equipment. These owners accumulate wealth; they accumulate capital. Enter the capitalists.

One consequence of specialization is growth in production. Another is increased exchange of goods. Some people produce machines that are used to manufacture shoes, while others specialize in producing clothing; and others again specialize in production of wine, or grain, or meat, or whatever. Thus, the individual household becomes less self-sufficient; what it produces becomes rather different from what it wishes to consume. An increasing need for an *exchange of goods* arises. It is most practical to organize this by introducing a monetary system, where *money* has the role of a *commonly accepted means of payment*.

To summarize, economic growth can be traced to two factors: (1) to increasing specialization in production, which leads to division of labour and development of new insights and skills; and (2) to investments in new machinery, with the accompanying accumulation of capital and development of new technology.

However, a third factor is also of importance—particularly in this book, which to a considerable degree is dedicated to a comparison of market and planned economies: namely, how the existing resources are utilized, in the short and long term, under the two systems. And it is precisely here that the decentralized decision-making system that the market consists of has outclassed the centrally governed system that a planned economy is based on.

In the remainder of Part I we will examine more closely how the existing resources are utilized better in a market economy than in a planned economy.

3
Supply and Demand

You can make even a parrot into a learned economist—all it must learn are the two words 'supply' and 'demand'.

Anonymous

SUPPLY and demand are central concepts in any economic system. By *supply* we mean how many units of a product are offered for sale at a given price. By *demand* we mean how many units of the product people wish to buy at this price.

3.1. Demand

As customers and consumers, we are confronted daily by decisions concerning what we should spend our money on. Most of us would like to acquire many things if we could only 'afford' them. Our wants are unlimited.

The woman who can afford only potatoes and bread would prefer meat and fish. The man who buys a Trabant would have bought a Lada, if his money had gone far enough. And the woman who buys a Lada wishes she could afford a Mercedes.

We thus face a *budget constraint*, in the sense that money can go so far and no further. We must therefore weigh the *usefulness* of each individual product or service against the usefulness of other goods and services.

As far as most goods are concerned, the usefulness or utility of the first unit is greater than the utility of the second, which again is greater than the utility of the third. When the household or individual increases consumption of a product, they finally approach a certain degree of 'saturation' with regard to consumption.

For example, if you already have two pairs of shoes and are considering buying a third pair, the need you have for the third pair will be weighed against alternative uses for the same money. Let us say the shoes cost 100 roubles. If a shirt also costs 100 roubles, and you decide that the utility of a new shirt will be greater than the utility of a third pair of shoes, you will come home with the shirt rather than the shoes.

However, should you be pleasantly surprised by the fact that the price

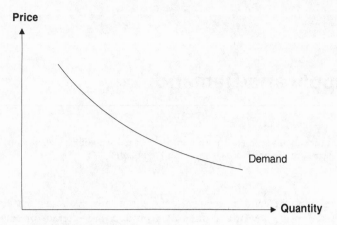

Fig. 3.1. The demand curve

When its price falls people wish to buy more units of a product. For this reason the demand curve slopes downward to the right.

of the shoes you were considering buying was not 100 roubles a pair but 90 roubles, it is quite possible that you will end up buying the shoes. The 10 roubles you then saved would justify your purchase of shoes.

We all think like this; it is only human nature. And if we take everyone into account, that is to say if we look at the *total demand* in society, the desire to purchase shoes and shirts and most other things will rise when the price of the product falls. Figure 3.1 shows the connection between price and quantity demanded. Along the horizontal axis the *quantity demanded* is measured, and along the vertical axis the *price*. A lower price will lead to increased demand, and the curve will fall to the right. This is true whether one lives under a system of economic planning or under a market-based system.

3.2. Supply in a Market Economy

When it comes to supply, the difference between the two systems becomes obvious. By 'supplied quantity' of shoes, shirts, etc., we mean the quantity of goods that producers supply for sale. In a market economy, where the individual producer aims at maximizing profits, it is clear that he will wish to increase production when the price rises. If the shoe manufacturer can sell his goods for 100 roubles a pair, he will produce 1,000 pairs per month. If he thinks that a pair of shoes can be sold for 115 roubles, he may wish to increase production to 1,200 pairs per month. Conversely, if his assessments indicate that 80 roubles a pair is what he can expect to get, he will find it profitable to reduce monthly production to 750 pairs.

If, in this story, we can imagine that there are ten or so shoe manufacturers who are all assessing the market, and we then add them together, we get the *market supply curve*, as shown in Fig. 3.2. The axes have the same designations as in Fig. 3.1. Whereas the demand curve falls to the right in a price–quantity diagram (Fig. 3.1), the supply curve rises (Fig. 3.2). A higher price provides an incentive to increase production. On the other hand, a higher price has a dampening effect on total demand.

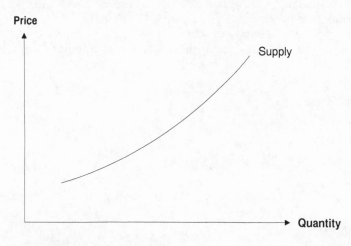

FIG. 3.2. The supply curve in a market economy

Note: When the price rises, existing producers wish to increase production—to reap greater profits. New producers may also be tempted to start up. For both of these reasons, the supply curve slopes upward.

3.3. Supply in a Planned Economy

In a planned economy the central authority decides how many pairs of shoes are to be produced, and also how many pairs each of the various shoe-making enterprises should produce. Further, the central authority must ensure that each shoe factory receives appropriate quantities of raw materials and semi-finished products. The overall plan, then, must be so detailed that each enterprise will, at the right time, in the right quantity, and in the right quality, receive the amount of inputs that are required. Only then will the enterprise be able to fulfil its role in producing the quantity stipulated by the plan. In addition, the central authority must decide on any purchase of new machinery, and on when and whether worn-out machinery should be replaced; that is to say, it decides what investments must be made.

If our shoe manufacturer is required to make 1,000 pair of shoes per month, he must therefore have enough leather, thread, plastic, rubber, buckles, etc., to make such production possible. These materials must be delivered by other enterprises. Since the central authorities will often have a planning horizon of one year, it will be difficult to change the planned figures during this period. This problem will be exacerbated by the fact that a change in the planned figures for the shoe factory would result in the need to change the planned figures for everyone who supplied raw materials to this enterprise.

In a planned economy, therefore, it is natural to expect the supply curve to be vertical. It is extremely difficult for an enterprise to increase production in the short term (see Fig. 3.3).

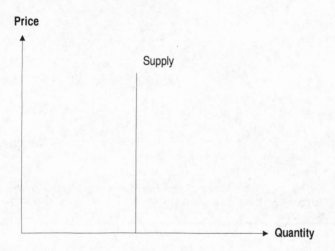

Fig. 3.3. The supply curve in a planned economy

Note: The plan stipulates total production for the period. This explicitly planned figure remains firm, independently of the price of the product. The supply curve in a planned economy is therefore vertical.

3.4. Shortages

Despite its striking simplicity, the above account of demand and supply can be used without further ado to throw light on some important economic and social problems, old and new. These applications serve two main purposes: they help to explain the persistence of some serious problems that plague both market systems and centrally planned economies to various degrees, and they point the way towards solutions to these problems, which will be discussed in greater detail in later chapters.

Consider housing first. Shelter remains among the prime necessities of man, for obvious reasons. Yet, in many market economies where food and clothing are plentiful, housing remains in short supply, as evidenced, for example, by the long waiting lists for public housing in many Western European countries. In planned economies, chronic housing shortages are legend. Why? The problem has do with demand and supply. A 'shortage' means that supply falls short of demand; hence the name. But we have seen that both supply and demand depend on price. This has a simple and fundamental consequence: if the price were right, there would be no shortage.

A typical rental housing market is depicted in Fig. 3.4. Along the vertical axis we have the price of housing, that is, the monthly rent. The rent charged depends on the size, standard, and location of the dwelling in question, among other things. Along the horizontal axis we measure the quantity of rental housing transacted in the market.

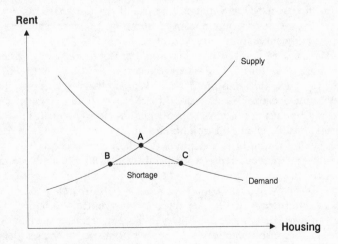

FIG. 3.4. The rental housing market

The demand curve slopes down just as in Fig. 3.1, because a decrease in rent increases the demand for housing, other things being equal. Families feel they can afford an extra room at lower rent, youngsters living with their parents free of charge feel they can afford to move out because the rent in the market has fallen, and so on.

The supply curve slopes up as in Fig. 3.2, because an increase in rent makes it profitable to construct new rental housing, and also to convert existing housing from other uses to the rental market. At point A, where the two curves intersect in Fig. 3.4, there will be no shortage: the quantity demanded is exactly matched by the quantity supplied. The price corresponding to the point of intersection is the 'right' price in the sense that, at this particular price, there will be no shortage because demand and supply are equal.

Reality in the market for housing, however, is generally better described by the gap between points B and C in the figure. Suppose the prevailing rent in the market is described by the vertical distance between the horizontal axis in the figure and points B and C. Point B then describes the quantity of housing supplied at this rent, which is lower than the rent at the point of intersection; at lower rent, supply of housing is more limited. Point C, on the other hand, describes the quantity of rental housing demanded at the lower rent; at this point, demand is greater than at the intersection point. Therefore, demand outstrips supply—which is precisely what is meant by a shortage. This result will come about as long as the prevailing market rent remains below the point where the two curves meet in the figure. This would still be true if the upward-sloping supply curve for a market economy borrowed from Fig. 3.2 were replaced by the vertical supply curve for a planned economy as in Fig. 3.3.

This discussion raises at least two important questions about efficiency and fairness. First, if housing shortages can be eliminated simply by raising the rent, why does this not happen? Why do housing shortages persist decade after decade in some countries? Second, if the problem can be wiped out by raising the rent, which would obviously benefit the owners of rental housing, would this solution be fair to those who pay the rent?

The answer to the first question is that there are forces at work in the economy that prevent rents from being raised. In planned economies rents are kept very low by decree, officially with the intention of benefiting tenants. In many market economies, rents are also maintained below the point of intersection in Fig. 3.4 through rent controls designed for the same purpose. In both cases, however, keeping rents artificially low is inefficient, because it reduces the supply of housing, increases demand, and thus creates a shortage.

As to fairness, artificially low rents do not benefit those who are excluded from the housing market because of limited supply and excessive demand. And for many of those lucky or privileged enough to get housing under difficult circumstances, low rents may be a mixed blessing, because low-rent housing tends to be poorly constructed in the first place and to deteriorate over time because of inadequate maintenance. It is no coincidence that housing standards in centrally planned economies are notoriously poor compared with those of market economies. But even in many market economies, especially in the United States, large segments of some urban areas have been devastated by inadequate upkeep and by the ensuing deterioration of residential housing, primarily because of controls that keep rents and housing prices permanently below the levels that would equate demand with supply and ensure normal maintenance and renewal. In the next chapter we will see how the removal of such legal restrictions would help solve this problem by allowing the rental market to gravitate automatically towards the point of intersection in Fig. 3.4.

3.5. Rationing of Credit and of Foreign Exchange

A similar story can be told of some other important markets in modern economies. The root cause of chronic credit shortages in many developing countries, for example, is that the price of credit—that is, the rate of interest—is kept artificially low by the authorities.

Consider Fig. 3.4 again. By relabelling the axes so that the vertical axis now describes the rate of interest and the horizontal axis refers to the amount of credit demanded and supplied in the market, we can apply the preceding analysis of the housing market to the market for credit, *mutatis mutandis*. In short, credit shortages could be eliminated by allowing interest rates to rise to the point where demand is matched by supply. This efficient outcome, however, is made impossible in many countries by legislation or by economic policies intended to help farmers and others whose livelihood depends on borrowed funds to be repaid at the end of the harvest.

The problem here is that interest rate controls (and sometimes also ill-designed usury laws), like rent controls, restrict the availability of credit and necessitate rationing, which frequently breeds discrimination and corruption as well as inefficiency. Sometimes this result is actually intended by public officials who are attracted by the discretionary power invested in them through the rationing of scarce credit to competing customers. The experience of many developing countries over many years indicates clearly that, by reducing the supply of loanable funds, interest rate controls that were intended to stimulate investment and foster economic growth have frequently had the opposite effect, and thus have defeated their purpose.

A very similar story can be told about the market for foreign exchange, to take a third example of the same kind. In countries where the price of foreign currency—that is, the exchange rate—is kept artificially low by the authorities for the benefit of those who like to pay as little as possible for foreign exchange, a shortage of foreign exchange will inevitably result, often with grave consequences for the country's ability to pay for imports of necessary goods and services from abroad and to honour its foreign debt obligations. The reason for the shortage is the same as in the examples of housing and credit given above: if the price of foreign currency is kept too low, supply will fall short of demand.

This simple but powerful idea can be applied to many other markets throughout the world economy. Indeed, the main reason for the chronic shortages of many goods and services that are the hallmark of centrally planned economies is that the prices in question are kept too low by the central planners. This is why price reform is essential to eliminate shortages in these economies.

But how should prices be reformed? Should the central planners simply raise prices to the point where supply has caught up with demand? Or are other more efficient means to the same ends available? We discuss these questions in the next chapter. And in the final chapter of the book, the need for price reform is analysed in a broader context.

3.6. Unemployment

Before concluding this chapter, it may be useful to take one more example of shortage—the shortage of work. Joblessness has been a recurrent problem in many parts of the world for a long time, with serious and sometimes devastating economic and social consequences for individuals and communities affected by it. This phenomenon too has do to with supply and demand.

In market economies, labour services, like commodities, are bought and sold in the market-place. Firms need labour to produce goods and services. The demand for labour by firms depends on the price of labour—on wages, that is. The higher the wages that firms must pay, the less labour they will want to employ, other things being equal. Workers supply labour to firms. Usually, the higher the wages workers can get, the more labour they are willing to supply. (This is not necessarily so, however, because an increase in wages raises the incomes of workers who may respond by working less in order to be able to enjoy more leisure. We bypass this possibility here to keep the story as simple as possible.)

The demand and supply curves that we have described above are drawn in Fig. 3.5. Here wages appear on the vertical axis and the quantity of labour services bought and sold (i.e. the number of hours of work) appears on the horizontal axis. The labour demand curve slopes down, as demand curves almost always do, and the labour supply curve slopes up, because higher wages encourage workers to work more.

Just as in our earlier examples, the wage corresponding to the intersection between the demand curve and the supply curve at point A in the figure is the 'right' wage, in the sense that at this wage demand and sup-

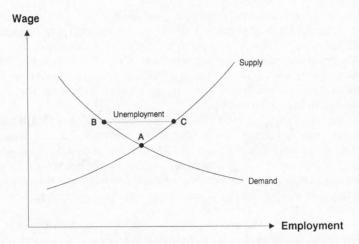

FIG. 3.5. Supply and demand in the labour market

ply are equal: all who want to work at this wage are able to find jobs.

In practice, however, the situation in labour markets is often better described by the gap between points B and C in Fig. 3.5. Suppose the prevailing wage in the market is described by the vertical distance from the horizontal axis in the figure to points B and C. Point B then describes the number of hours of labour demanded by firms at this wage, which is higher than the wage at the point of intersection; at a higher wage, there is less demand for labour. Point C, on the other hand, describes the number of hours of labour supplied by individuals at the higher wage; at this point, supply is greater than at the point of intersection. The supply of labour exceeds the demand for labour—which is precisely what is meant by unemployment. This result will come about as long as the prevailing wage in the market remains above the point where the two curves meet in the figure. This would still be true even if the upward-sloping supply curve were replaced by a vertical supply curve (which would be appropriate for the case where employees do not increase their work effort in response to a wage hike because of their interest in undiminished leisure).

But why do wages in the real world remain above the point of intersection shown in the figure? What keeps them there? This is a crucial problem in many market economies. To answer this question, it may help to think about who gains and who loses from a *status quo* characterized by unemployment. Those who have jobs at point B in the figure are clearly better off than they would be at the full-employment point A: they have an incentive to prevent wages from falling from B to A. On the other hand, those who are unemployed at point B, but would be able to find jobs at the lower wage at point A, are worse off: the high wage at point B does them no good because their wage is zero. (To be precise, they normally receive an unemployment benefit which is lower than their wage would be at point A.) Thus, there is a conflict of interest between workers and their unions: the 'haves' want wages to remain high, while the 'have-nots' would like to see them fall.

What about employers? At first glance it would seem that their interests would normally coincide with those of the 'have-nots' in this case, but this is not necessarily so. The reason has to do with the quality of labour services. Employers may be reluctant to press wages down because they may fear that this could demoralize workers: the best workers might leave for better-paid jobs elsewhere and the remaining workers might not be motivated to perform as well as before. Therefore, employers may prefer the status quo with the higher wage. In sum, the power structure in the labour market may explain, to some extent at least, why wages sometimes are kept at a level that is inconsistent with jobs for all. We shall return to labour market problems in future chapters.

4

Economic Adjustment

> We might as reasonably dispute whether it is the upper or the under blade of a pair of scissors that cuts a piece of paper, as whether the value is governed by utility (demand) or cost of production (supply).
>
> Alfred Marshall

IN this chapter we study how economic adjustments take place in a market economy. To this end we put Figs. 3.1 and 3.2 together in a diagram (Fig. 4.1), just as we did in our example of the housing market in the preceding chapter. In order to gain a better understanding of what is happening, we have placed specific numbers along the axes. We have designed this example in such a way that the supply and demand curves intersect at point A where the price of shoes is 100 roubles per pair, and the quantity produced (and sold) is 20,000 pairs of shoes per month.

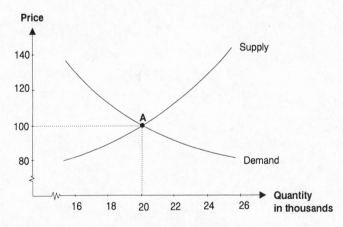

FIG. 4.1. Supply and demand in a market economy

Note: At point A the market is cleared. The producers manage to sell precisely their total production (20,000 pairs of shoes per month) for the price they had expected (100 roubles per pair). At the same time, everyone who wants to can buy shoes for 100 roubles a pair. At point A the market for shoes is in equilibrium.

4.1. Market-Clearing

At point A we have the unique situation that all those who wish to purchase shoes can buy what they want for 100 roubles a pair. And all those who wish to produce and sell shoes can sell precisely the amount produced for 100 roubles a pair.

At point A we say the market is cleared. A particularly important property of the market economy is that it arrives at this point by itself. It is, as Adam Smith put it, as if 'an invisible hand' steers producers and consumers to this point. There is no need for the economy to be directed by a central authority; nor is there any need for an overall plan of production.

This important point requires a more precise explanation. In Fig. 4.2 we have again drawn in the two curves from Figs. 3.1 and 3.2. The problem facing the individual shoe manufacturer is that he does not know how much his competitors intend to produce. He knows his own supply curve but not the supply curve for the entire market. Further, he lacks reliable information concerning the connection between price and quantity demanded; that is to say, the market's demand curve is also unknown to him.

Let us now suppose that all of the shoe manufacturers base their calculations on an expected selling price of 130 roubles a pair (point B on the supply curve). Each producer will calculate his most profitable production

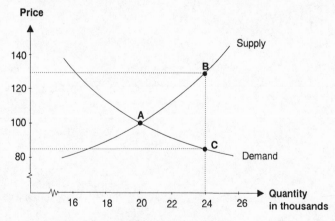

FIG. 4.2. Supply exceeds demand in a market economy

Note: As a point of departure the producers had planned on a price of 130 roubles per pair of shoes. Based on such an estimate, they supply a total of 24,000 pairs (point B on the supply curve). But for people to be willing to buy so many pairs of shoes, the price must go down to 85 roubles a pair (point C on the demand curve).

based on this price. The total supply of shoes in this month will therefore be 24,000 pairs.

However, the producers have miscalculated. The utility that people have for shoes is not so great that they want to buy as many as 24,000 pairs at 130 roubles a pair. If, on the other hand, the price is reduced to 85 roubles a pair, the whole production will be sold (point C on the demand curve).

Having learned from their mistake, the producers choose to reduce production in the following month. They now supply a total of 19,000 shoes (see Fig. 4.3), and expect to obtain a price of 95 roubles a pair (point D on the supply curve). But now they are pleasantly surprised. With only 19,000 pairs of shoes for sale (compared with 24,000 in the previous month), the producers will obtain a price of 105 roubles a pair (point E on the demand curve).

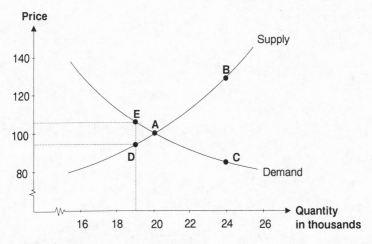

FIG. 4.3. The market approaches equilibrium

Note: With production of 19,000 pairs of shoes the producers can obtain 105 roubles a pair, compared with the expected price of only 95 roubles. At the same time, we see that the market for shoes has approached the state of equilibrium marked by the intersection between the supply and demand curves (point A).

We can guess the rest of the story. The following month there will be a slight increase in production. Through further 'trial and error' the market will approach point A, where the price is 100 roubles for one pair of shoes, and where 20,000 pairs of shoes are produced and sold each month. When the market has reached this point it will come to rest. Point

A is thus the *state of equilibrium* of the shoe market. At this point the market is stabilized. This means that the total production of the shoe manufacturers will be sold. Further, at this point there will be an absence of queues and no shortage of goods. Everyone who wants and can afford a pair of new shoes for 100 roubles per pair will have their demand satisfied. Those who would like to buy a pair of new shoes but feel that 100 roubles is too expensive will choose to spend their money on something else.

4.2. Price-Setting in a Planned Economy

If we now place Figs. 3.1 and 3.3 in the same diagram, the result will be Fig. 4.4, which gives us a picture of the situation in a planned economy. The difference between a planned and a market economy concerns the supply curve: here it is vertical; in a market economy it was upward-sloping.

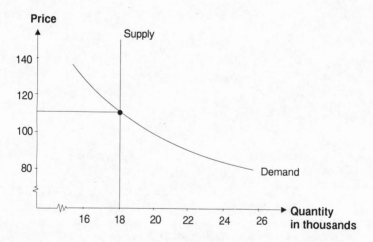

FIG 4.4. Supply and demand in a planned economy

Note: The demand curve is as in Fig. 4.3. However, the supply curve is vertical at a quantity of 18,000 pairs of shoes. This is due to the fact that in the overall plan for the country's economy the authorities have determined that 18,000 pairs of shoes are to be produced per month.

There is another and very important difference between a planned and a market economy: in addition to determining the *quantity produced*, the authorities, or the body of central planning, must also determine the *price* to the consumer.

We can now imagine three situations illustrating the workings of a planned economy, all based on the fact that the central authority has decided that 18,000 pairs of shoes are to be produced per month. In the first situation (see Fig. 4.5) the authorities determine that the price to consumers shall be 130 roubles a pair. But at this price consumers demand only 16,000 pairs of shoes (point F on the demand curve). Thus, the enterprises have a surplus of 2,000 unsold pairs.

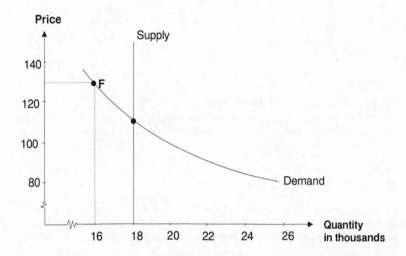

Fig. 4.5. Supply exceeds demand in a planned economy

Note: At a price of 130 roubles for a pair of shoes, only 16,000 of the 18,000 pairs produced will be sold: point F on the demand curve. This implies an unintended increase in inventories of 2,000 pairs of shoes.

In the second situation (see Fig. 4.6) the price is stipulated at 90 roubles a pair, and the same quantity of 18,000 pairs of shoes is produced. At 90 roubles a pair, the total demand will be 22,000 pairs (point G on the demand curve). But since only 18,000 pairs are available, some people will have to go without new shoes. Instead of increasing inventories of unsold shoes, the situation will be characterized by a lack of goods and by queues. If those consumers who are lucky enough to buy a pair of shoes queue up for an average of a quarter of an hour, we can say that the price they pay is 90 roubles plus 15 minutes.

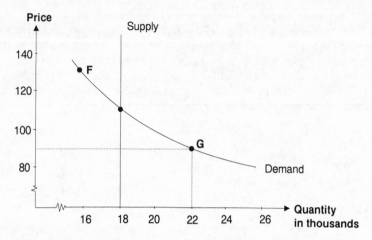

Fig. 4.6. Demand exceeds supply in a planned economy

Note: At a price of 90 roubles per pair, people will wish to buy 22,000 pairs of shoes altogether: point G on the demand curve. But since only 18,000 pairs are supplied, there will be a shortage of 4,000 pairs.

SUPPLEMENT 4.1. LOVE, MOSCOW-STYLE

I met my husband, Ivan, in 1984. In the queue. We were lining up for washable wallpaper. I liked the man right away. He was witty. He was no. 1984 in the line—he said he was keeping up with the times. He was punctual. Ours was not a one-day queue, so we met in line on Wednesdays at 7 p.m.; Ivan was always on time. He was kind. Ivan explained to whoever asked what we were queuing up for. We became friends. We went to the theatre together. I was ahead in the line for Bolshoi tickets: my number was only 2245, his was 5894. We began going to the theatre often.

A couple of months passed and Ivan led me to the marriage registration office. There was a very short queue there! We were the sixteenth couple.

Years passed. We live well. My husband is in the apartment queue, no. 3857463947618. So at the moment we stay with my mother. It is a good room, close to downtown. We don't have a telephone yet, but we've got the number: 959–14–94. That's not the phone number, it's the queue number.

We have a son, Vasili. A clever boy he is! He can even write his queue number for kindergarten entry: 96.

Svetlana Vozlinskaya
Moscow
(*International Herald Tribune*, 6 November 1991)

The third situation will get it just right. In Fig. 4.7 the authorities have determined a price for shoes of 110 roubles per pair. When production lies at 18,000 pairs per month, the result will be a clearing of the market (point H) and everyone will be satisfied.

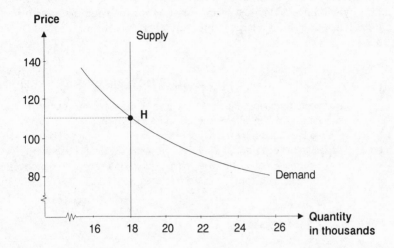

FIG. 4.7. Supply equals demand in a planned economy

Note: Given a monthly production of 18,000 pairs of shoes, we see that precisely this quantity will be demanded when the price is set at 110 roubles a pair. We achieve equilibrium in a planned economy at point H.

4.3. Information Problems under Planning

But—and this is the big question—how can the planners manage to control the economy (in this case the market for shoes) so that the situation in Fig. 4.7 arises? In principle—i.e. in purely mathematical terms—it has been demonstrated that this is possible. What is needed is a mechanism for *feedback* to the body of central planning from the shoe shops. When the goods pack the shelves, and the central authority is informed of this, a message is sent back that the prices must be reduced. When the opposite happens, and there are regular queues outside the shops, then the prices must be raised. By using such a mechanism, the planners can ensure that adjustments take place via changes in *prices*.

Alternatively, or in addition, the central planning body can revise its production plans, i.e. can order adjustment via changes in *quantity*. When the point of departure is 18,000 pairs of shoes at 130 roubles (Fig. 4.5) and 2,000 pairs remain unsold, an attempt can be made to remedy the situation the following month by reducing production, for example by 1,500 pairs of shoes. The problem posed by such a form of adjustment is that all orders to shoe factories of leather, rubber, thread, etc., must then be changed. Changes in the supplies of the finished product will thus create a chain reaction backwards in the production apparatus. The central authority will have to lay new plans not only for the shoe-manufacturing enterprises, but also for a large number of other enterprises.

SUPPLEMENT 4.2. MILLIONS OF GOODS, BILLIONS OF ORDERS

Taking a comprehensive look at the whole Soviet economy, there were an estimated 24 million different goods and services. Production of these required 5 billion different types of orders. In practice it proved completely impossible to keep track of all of this. The result was that changes in explicit planning figures for what the individual enterprise should produce, based on feedback from shops, were far from common.

Source: After Aganbegyan (1988).

The planned economy suffers from a lack of flexibility. The central authority is required to solve problems that have been shown in practice to be impossible to solve. The result is that market-clearing, as illustrated in Fig. 4.7, is far from guaranteed. Experience would seem rather to indicate that the opposite is true.

4.4. Self-Regulatory Mechanisms in a Market Economy

The strength of the market economy lies in its inherent *self-regulatory* mechanisms (the 'invisible hand'). When producers realize that their goods can no longer be sold at the existing prices, they will utilize both adjustment mechanisms on their own initiative; that is to say, they will reduce both the quantity produced and the price. The reason for such behaviour lies in the profit motive. If producers do not listen to the signals given by the market, they will be outcompeted and may lose their capital base.

This brings us to another and very important point as far as the functioning of a market economy is concerned: namely, the significance of competition for keeping down production costs. When there are several alternative suppliers of a product, and everyone wishes to achieve the highest possible profit, a struggle to capture customers will ensue. This struggle is the essence of competition.

If, for example, one of our shoe producers in the example above finds a cheaper way of producing shoes, he will be able to increase his profit by reducing the price slightly, thus gaining a greater share of the market. The other enterprises, his competitors, will experience a falling-off of their sales and profits. They will have to make an effort to develop cheaper methods of production. Those who do not succeed in doing this will be unable to compete and will have to close down their enterprises.

A particularly important consequence of competition is that enterprises become open to *innovation*. This leads to research and development, aimed at utilization of new methods of production, new raw materials, new ways of organizing production and distribution, etc. Viewed in a longer perspective, the market's capacity to innovate is of very great importance in explaining the differences today in technological level and living standards between market economies and planned economies.

Let us once again use the shoe trade as an example. Those enterprises that realized early that synthetic rubber was a cheaper and better material for some types of shoe-soles gained a lot by replacing leather with rubber. Other shoe-manufacturers, observing that they were becoming less competitive, were forced to follow their lead. Each of them must strive to keep up with product development and innovation in order to survive financially.

In the same way as consumers in a market economy characterized by competition are faced by alternative producers, it is important for producers themselves to have alternative suppliers to choose from. Several tanneries that supply leather, several thread factories that supply laces and thread, etc., will encourage competition further down the chain of production. What applies to the shoe-manufacturing enterprises will also apply to the tanneries and everyone else; driven by the desire for maximum profit, they will all make a great effort to remain competitive. And it is precisely this effort that results in an effective utilization of resources.

If we think in more general terms, most forms of production require three types of inputs: labour, capital (i.e. machinery, buildings, etc.), and raw materials. The producer, who has a vested interest in the lowest possible costs and highest possible profit, will change the composition of the inputs if their prices are changed. If wages rise, this will act as an incentive towards increased use of machinery in production. Instead of having 100 workers and 10 machines, it may now be profitable to acquire 2 more machines and reduce the work-force to 90.

Or if the rate of interest on bank loans rises, that is if it costs the enterprise-owner more to borrow money, then the producer may employ more workers rather than borrow money to replace a worn-out machine.

4.5. Consumers in Focus

In a market economy the capitalist or enterprise owner is in command. He decides what is to be produced, how much, and how. All the time he has his eye on one prime goal: to earn as much money as possible for himself—or put another way, to maximize the profit on his own invested capital.

This does not appear to be a particularly noble objective. The greedy capitalist is certainly not a sympathetic figure in the popular imagination. But this does not imply that it does not make sense to give the profit motive plenty of scope when it comes to solving the economic problems in society. For behind the capitalist we will find the consumer; and it is only by satisfying the needs of consumers, as they are expressed in the demand for goods and services, that the capitalist can earn money. Thus, we can say that it is the consumers' interests that are in focus in a market economy. The structure of production and distribution are ultimately determined by consumers in a market system. Producers serve consumers.

In a planned economy, by contrast, it is the producers' interests that prevail. Producers receive orders from central planners without having any direct contact with consumers. Therefore, consumer interests tend to be neglected in planned economies. We will return to this problem among others in Chapter 6. The market economy also has its weaknesses; Chapter 9 is devoted to the discussion of some of these.

4.6. 'Market Socialism'

Before concluding this chapter, we will briefly consider the extent to which the principles of the market economy can be utilized in an economy in which the state owns the capital equipment. Is it possible to achieve effective production by utilizing the market, and an even distribution of income by utilizing planning?

If we take a broad view, we can distinguish between two forms of ownership (private and state) and two forms of co-ordination of the economic decisions in a society (planning and market). This results in four possible combinations, as shown in Table 4.1.

TABLE 4.1.   Four ways of organizing an economy

	Private ownership of the means of production	State ownership of the means of production
Co-ordination by market	1	2
Co-ordination by planning	3	4

The Hungarian Professor Janos Kornai, who has been working since the 1950s on questions concerning economic planning at the same time as he has been engaged by some of the foremost American universities (such as Stanford and Harvard), has concluded that the choice stands between alternatives 1 and 4 in the table. No other alternative to either capitalism (1) or traditional socialism (4) has functioned particularly well.

Alternative 2, where the state owns the means of production while the market is utilized for co-ordination of decisions, has been given the name 'market socialism'. Variants of this alternative have been tried out both in Hungary and in Yugoslavia. The idea behind 'market socialism' is to free the state-owned enterprises from a central plan, and instead to give them price signals they must adjust to. When planners no longer stipulate explicit target figures for production, but instead simply determine prices for all sorts of things, and let enterprises freely adjust to these, a market economy is simulated without actually being introduced. The purpose is to obtain the effectiveness from the system that a market economy has shown it can achieve, while at the same time determining the distribution of income centrally. Instead of Adam Smith's *invisible* hand, then, you introduce an *artificial* hand.

However, this artificial hand has not functioned according to intention. This is due to at least three factors. First, the planners retain power as long as the state retains ownership of the means of production. Direct control through explicit target figures is replaced by control through administratively stipulated prices, various taxes and subsidization arrangements, etc. The framework within which the individual enterprise must function is influenced by the decisions the bureaucrats or planners are obliged to make. It will thus be more important for enterprise leaders to maintain good contact with the bureaucrats and to provide them with appropriate information, which will make life comfortable for the managers of the firms, rather than ensuring effective production ('deceiving without lying').

Second, enterprises often regard themselves as best served by having the state as owner; that way they can avoid the danger of bankruptcy. Surveys have revealed that Hungarian state undertakings that earn good money *before* paying direct and indirect taxes have approximately the same profitability as the less effective enterprises, as calculated *after* direct and indirect taxes. This is quite simply because the good enterprises are taxed and the poor ones are subsidized. Another example comes from Poland. In the electro-mechanics industry, profits in 1988 were reduced by taxation from 16 to 7 per cent, while losses in the foodstuffs industry of 9 per cent were converted with the help of subsidies to the same profit level of 7 per cent. Corresponding figures in a market economy would inevitably have led to bankruptcies among a number of producers of foodstuffs.

The third reason why 'market socialism' does not work lies in the lack of incentives. When distribution of income has little to do with the profits resulting from production, it will be difficult to motivate leaders and workers to make maximum effort. The incentive to work hard is weaker, and day-to-day improvements in organization and production techniques will remain absent, when there are no incentives for such improvements. The conflict between effective production and an even distribution of income has not been possible to solve by using alternative 2 in Table 4.1.

4.7. A Centrally Planned Market Economy

The final alternative in Table 4.1, private ownership and co-ordination through planning (alternative 3), would be thought by some to be illustrated by experience with indicative planning in France. In France (and also in other Western countries), forecasts for industrial development are prepared; then, through economic policy, an attempt is made to fulfil these forecasts. However, the public servants working on economic policy have scarcely any independent power over individual firms. Rather, it is a question of general public-sector support for research and development (first and foremost at universities and technical colleges), more favourable tax conditions in connection with location of production in poorly developed areas, particular concentration on selected areas (such as 'high-tech'), etc.

As long as public authorities do not have the right to interfere with the operations of the individual enterprise, co-ordination of decisions will take place not through planning but through the market: we are back then to alternative 1 in Table 4.1. The authorities' contribution will be limited to altering the general business environment so that, in their own interests, enterprises will conduct their operations in line with the wishes of the authorities. Indicative 'planning' in France and other market-based economies thus has no similarity to production planning of the traditional Central and Eastern European type.

SUPPLEMENT 4.3. WHAT ABOUT THE 'SCANDINAVIAN MODEL'?

There is a widespread belief that the Scandinavian countries (Denmark, Norway, and Sweden) have found a 'third way', where the good properties of a market economy are combined with extensive public-sector influence in the economy.

However, this 'mix' does not take the form of a centralized

plan for the most important decisions, and a market economy for all others. Two factors characterize the Scandinavian model.

First, the business sector consists mainly of privately-owned enterprises. These enterprises operate under market-economy conditions. On the domestic market they compete with imports from abroad. This leads to effective production and a good ability to pay high wages.

Second, the earned income of employees and the profits of enterprises are taxed harder than in other industrialized countries. These tax revenues are utilized among other things for production of health and education services by the public sector. These services are again supplied to the individual inhabitant free of charge or at a symbolic price. Moreover, there is a comprehensive publicly financed welfare system, which benefits, among others, families with children, the unemployed, and pensioners.

Thus, the Scandinavian model does not consist of state-owned enterprises whose production of goods is controlled by an overall plan. However, through high direct and indirect taxes, a major part of the consumption of services has been socialized, compared for example with Thatcher's Britain.

The relevance of the Scandinavian model to the challenges now facing the countries of Central and Eastern Europe has been well grasped by the weekly newspaper *The Economist* (13 January 1990):

> The main debate in economic reform (in Eastern Europe) should therefore be about the means of transition, not the ends. Eastern Europe will still argue over the ends: for example, whether to aim for Swedish-style social democracy or Thatcherite liberalism. But that can wait. Sweden and Britain alike have nearly complete private ownership, private financial markets and active labour markets. Eastern Europe today has none of these institutions; for it the alternative models of Western Europe are almost identical.

In summing up, Professor Janos Kornai concludes that, if a society wants to establish a market economy, it will be necessary for it to accept private ownership. Furthermore, when the state owns the means of production, it will not be possible to avoid a form of centrally controlled economy, where the planners have considerable power. Despite extensive attempts at 'market socialism', especially in Hungary, Yugoslavia, and China (in agriculture), a viable 'third way' has not yet been found.

4.8. 'Economic Man'

In what follows we will base our discussion on the premiss that the transition from plan to market will necessitate private ownership of the means of production. An exhaustive discussion of all imaginable variants of the 'third way', such as worker-controlled enterprises, various co-operative schemes, and the like, will therefore be irrelevant to the subject matter we are concerned with in this book.

Instead, we will introduce the somewhat uncharming 'Economic Man' (*Homo oeconomicus*). He is a rational decision-maker who tries to achieve as much as possible with as little effort as possible. Moreover, he prefers to avoid taking risks, but is not unwilling to do so if the expected reward is great enough. 'Economic Man' is fully informed as to prices and technology. Furthermore, he knows what he likes and does not like. This means that he is perfectly capable of making all the choices that his economic adjustment requires.

'Economic Man' is well known from textbooks on economics. The disadvantage is that few people really feel able to identify with this person. Human motivations and interests are far more complex and varied, and our behaviour is rational in a way other than that attributed to 'Economic Man'. It is obvious that most normal people are interested in things beyond their own material welfare. Our actions and decisions are seldom determined solely by economic considerations. In order to enlarge our understanding of economic problems, it can nevertheless be useful to isolate those aspects of human behaviour that have an economic basis.

The advantage of the simple caricature of 'Economic Man' is that it provides an uncluttered point of departure for an analysis of the way the economy functions. If we look at the way people as a whole behave, we will find that they often on average approach the behaviour of 'Economic Man', even though it may be difficult to find specific individuals who do so. If, for example, the tax on goods and services is raised by one percentage point, one can calculate beforehand and with a reasonable degree of accuracy by how much total consumption will decrease, how much total taxes will increase, etc., although it will normally be impossible to predict how any one individual will react. Some will react strongly, others not at all, and some will even do the opposite of what one would expect. The point is that extreme reactions tend to cancel one another out, so that the result we observe for the economy as a whole resembles the behaviour of our 'Economic Man'. The insight to be gained from analyses based on 'Economic Man' has been shown to be extremely useful in the work of putting together a reasonably effective market economy.

To lay the groundwork for a more detailed discussion in Part II of how a market economy works, in the next chapter we shall apply the demand and supply apparatus developed in this chapter and the preceding one to study the functioning of a market economy—not just individual markets, but the economy as a whole. In the following chapter we shall then look more closely at the way a planned economy functions. A solid understanding of the system one wishes to get rid of will clarify many of the problems encountered in the transition to the new system, including the need for devising sound economic policies and creating conditions for a healthy business environment.

5

The Interplay of Aggregate Supply and Demand

The outstanding faults of the economic society in which we live are its failure to provide for full employment and its arbitrary and inequitable distribution of wealth and incomes.

John Maynard Keynes

IN different walks of life it is useful to distinguish the forest from the trees. This is true of economics. Individual firms, industries, and markets can be fascinating to observe; but this must not obstruct our perspective on the national economy as a whole—or on the world economy, for that matter.

5.1. Micro- and Macroeconomics

Economists have divided their discipline into two major branches. *Microeconomics* is the economics of individual consumer behaviour and the behaviour of the firm. It deals with individual markets (for shoes, shirts, and so on) and market structures, and their interrelations. *Macroeconomics*, on the other hand, is concerned with the structure and functioning of the economy as a whole rather than its individual parts. The study of profit-seeking behaviour of individual enterprises belongs to microeconomics; the aggregate investment resulting from the individual investment decisions of all firms and households in the economy is within the purview of macro-economics. To take another example, the market for shoes in Gdansk is a microeconomic concern, but the market for all goods and services in Poland—shoes, shirts, and all the rest—belongs to macroeconomics.

This having been said, the boundary between microeconomics and macroeconomics is not always very sharp; there are large areas where the two overlap. Moreover, the two branches are intimately intertwined. After all, macroeconomic quantities such as aggregate supply are simply the sum of corresponding microeconomic quantities supplied by individual firms. Nevertheless, the distinction between microeconomics and macroeconomics is a useful way of organizing one's thoughts about economic problems and policies.

In microeconomics we study the determination of prices and quantities through the interplay of demand and supply in individual markets. Our example of the market for shoes in the preceding chapter was an exercise in microeconomics. Does that example tell us anything about prices in general or about the overall cost of living? No. Does it tell us anything about the total quantity of goods and services produced in the economy? No, of course not. To study the determination of the general price level and aggregate production in the economy as a whole, we need to enter the field of macroeconomics.

SUPPLEMENT 5.1. THE GENERAL PRICE LEVEL

In microeconomics we study, for example, how the price of shoes affects the demand for shoes and the supply of shoes, and how the price and quantity of shoes are determined by the interaction of demand and supply. In macroeconomics, when we study the interplay between *aggregate* demand and *aggregate* supply in a country, we must look beyond the price of shoes: we must study the prices of all goods and services that are bought and sold in the economy. To do this, we must introduce a new concept: the general price level, or just the price level for short. The price level can be found by computing how much it would cost to purchase a wide spectrum of goods and services on a given date, e.g. 1 January 1991. Then at a later date, say 1 January 1992, we compute how much it would cost to purchase exactly the same basket of goods and services. If we have to pay more in 1992 than in 1991, we say that the price level has risen.

Because the price level reflects the value of a broad spectrum of goods and services at different points of time, the prices of individual goods and services do not have to be unchanged even if the general price level does not move. The price level says nothing about changes in *relative prices*. Suppose, for example, that the prices of electricity and haircuts have increased, while the prices of tobacco and chicken have decreased and all other prices have remained the same. Tobacco and chicken then have become relatively less expensive than electricity and haircuts; and, since the prices of all other goods and services remain unchanged, tobacco and chicken have also become relatively less expensive than those. Electricity and haircuts, on the other hand, have become relatively more expensive than all other goods. The general price level will have risen only if the prices of electricity and haircuts have risen relatively more than the prices of tobacco and chicken have fallen.

In this chapter we show how the main ideas involved in the simple example of the market for shoes discussed in the preceding chapter can be applied to the analysis of the economic system as a whole. This will enable us to see how the general price level is determined, and also *gross national product (GNP)*—that is, the total of all goods and services produced in the economy, and hence the national income. Our interest in national income stems not only from the fact that it is the basis of our material welfare, but also from its intimate relationship to other things we care about, especially employment and unemployment. Next, we study the determinants of the general price level and of GNP more closely. These determinants include many things that are beyond human control (weather, for instance) and other things that are manageable to some extent, through the economic policies of the government. Finally, we consider the effects of various external shocks (such as the quadrupling of oil prices in 1973–4) and of economic policies on the price level, on national income, and on unemployment.

5.2. Aggregate Demand

We saw in earlier chapters how an increase in the price of shoes reduces the demand for shoes, other things being equal. Since the consumer has a limited budget, she usually buys less of things whose prices have risen, and more of things that have become relatively less expensive. But what can be said about the relationship between the general price level in the economy and aggregate demand for goods and services?

First, the terms must be given precise content. The *general price level* is a properly weighted average of the prices of all the goods and services produced in the economy. *Aggregate demand* is the total demand by households, firms, the government, and foreigners (who import our products) for goods and services produced in the economy. The price and quantity of shoes and shirts bought and sold will affect both the general price level and aggregate demand. These effects, however, will be minuscule because of the very large number of other goods and services produced and consumed.

SUPPLEMENT 5.2. NATIONAL INCOME ACCOUNTING

Gross national product (GNP) is an important indicator of the volume of economic activity in a country. Specifically, GNP indicates the total value of the goods and services produced in the national economy. But GNP can be defined equivalently in more ways than one: it also indicates the total level of national income: that is, the sum of the wage earnings of workers, interest earnings of capitalists, rents earned by land-owners, and so on.

Moreover, GNP can also be defined as the total expenditure of households, firms, government, and foreign nationals on domestically produced goods and services. By this definition, GNP equals the sum of the consumption expenditures of households (on food, haircuts, etc.), investment expenditures of firms (on machinery, equipment, etc.), government expenditures (teachers' salaries, hospital construction, etc.), and export earnings (i.e. foreigners' purchases of domestic exports), less imports of goods and services. Imports are subtracted to avoid double-counting because consumption expenditures, investment, government spending, and exports all include some foreign-produced goods and services which must be deducted to make sure that GNP is appropriately defined as the sum of expenditures on domestically produced goods and services. Hence one of the most frequently encountered equations in economics: GNP equals consumption (C) plus investment (I) plus government spending (G) plus exports (X) minus imports (M); in short,

$$GNP = C + I + G + X - M.$$

Suppose now that the general price level rises for some reason, other things being equal. This will affect aggregate demand in two important ways. First, the real value of our money will decline. Why? By making goods and services more expensive in general, an increase in the price level reduces the *purchasing power* of our money and of other financial assets whose face value has not changed. The result is that we can buy less goods and services with given nominal income; i.e., an increase in the price level reduces aggregate demand.

In the second place, an increase in the price level in our country will make domestically produced goods and services more expensive relative to foreign goods and services, other things (among these, the exchange rate) being equal again. This means that consumers at home will be tempted to buy less home-produced goods and services because they have become more expensive relative to foreign-made products. Therefore, imports will rise. This also means that producers at home will be able to sell less of their products abroad because they have become more expensive relative to competing products in foreign markets. So exports will fall. Thus, through foreign trade links, an increase in the domestic price level will reduce aggregate demand for home-made goods and services. The effect of the price level on aggregate demand through this channel can be crucial in countries that depend heavily on international trade.

In sum, by reducing the purchasing power of money, by increasing imports, and by decreasing exports, an increase in the general price level reduces aggregate demand for domestically produced goods and services. This relationship is described by a downward-sloping aggregate demand curve in Fig. 5.1. Notice the resemblance to Fig. 3.1 in Chapter 3: the only difference between the two figures lies in the labelling of the axes. Figure 3.1 was used to describe the demand curve for shoes; Fig. 5.1 describes the demand curve for shoes *and* everything else that is produced at home. This figure summarizes the demand side of the economy, under either a centrally planned economy or a free-market economy.

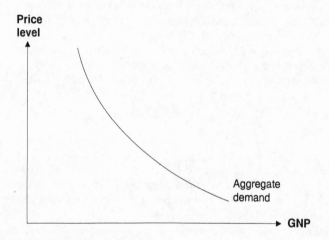

FIG. 5.1. The aggregate demand curve

Note: When the general price level rises, the purchasing power of people's money declines, so they wish to buy less goods and services. In addition, consumers to some extent shift their purchase of goods and services from domestic suppliers to foreign ones. Furthermore, at higher prices producers find it more difficult to sell their goods and services abroad so exports fall. For all three reasons, the aggregate demand curve slopes downward.

5.3. Aggregate Supply

In order to complete our macroeconomic picture of the market for goods and services, we also need to consider the supply side, just as in the microeconomic example of the market for shoes. Firms produce goods and services and supply them to the market. The production process requires the use of labour, capital, raw materials, and other inputs in some suitable combination which varies from product to product. Of these factors of production, human labour is generally the most important by far, in the sense that the lion's share (two-thirds is a common figure in market economies) of the cost of production is labour cost—i.e. wages. Therefore, to keep our story as short and simple as possible, let us for the sake of the argument concentrate on labour by regarding it as the sole factor of production in the economy.

Consider what happens on the supply side of the economy when the general price level rises for some reason, other things being equal. Firms now receive higher prices for their products, but they pay the same (nominal) wages to their workers as before because wages are held constant, by assumption (this is what is meant by 'other things being equal'). If the firm previously had to produce and sell 10 units of output to pay one worker for a day's labour, only 9.1 units are now required if the price of output increases by 10 per cent. Therefore, the *real* cost of labour to firms falls. The decline in real wage costs and the resulting increase in profits signal to the firms that they should employ more workers and expand production. This is the main channel through which an increase in the general price level tends to raise the aggregate supply of goods and services in the economy.

This relationship is described by an upward-sloping aggregate supply curve in Fig. 5.2. Again, notice the resemblance to Fig. 3.2 in Chapter 3. The only difference between the two figures is that one describes a single market, while the other covers the economy as a whole. Notice also that the above description of the aggregate supply curve applies only to market economies. In planned economies, the supply curve for each product is independent of price, as described in Fig. 3.3 in Chapter 3. If all goods and services are added up, the resulting supply curve also becomes vertical; i.e., the aggregate supply curve in a planned economy is independent of the general price level.

The aggregate supply curve shows the total amount of goods and services that firms want to produce at any given general level of prices and wages. The higher the price level for given wages, the more the firms are prepared to produce.

The above simplified description of the supply side of market economies is not without problems, however. The main problem is not that we have concentrated on labour and thus abstracted from capital and other relevant inputs into production to simplify the story: the main problem has to do with our assumption that wages remain fixed when the price level rises. Under this assumption, real wage costs fall, as we saw. This is good for the firms, but not so great for the employees. A higher price level implies a higher cost of living, and hence a lower purchasing power of wages. This means that the employees are worse off than before.

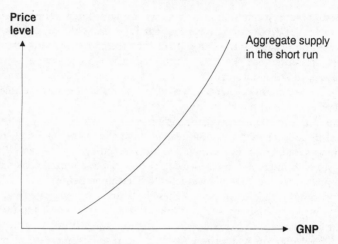

Fɪɢ. 5.2. The aggregate supply curve in a market economy

Note: When the general price level rises, the real wage cost of firms falls and profits rise if wages remain unchanged. Firms react by hiring more workers and expanding production. Therefore, the aggregate supply curve slopes upward for given wages.

Not all workers will be worse off, however. Remember that an increase in the price level encourages firms to produce more and to employ more workers. This means that previously unemployed workers get jobs. They, like the employers, are clearly better off with the higher price level.

Now the question is this: 'When the price level rises (and profits rise, too), will the workers—individually or through their labour unions—insist on higher wages?' The general answer is: 'Maybe, maybe not.' If workers do not insist on higher wages, the story as told above applies. On the other hand, if the workers do insist on a wage increase to compensate for the rise in the price level (and in profits), the story must be changed. To see how, imagine that the price level rises by 10 per cent and that workers insist on, and receive, a corresponding wage increase of 10 per cent. What has happened? The workers are just as well off as before because the purchasing power of their wages has not changed. Not only that, but their employers are also in the same position as before. Their real cost of production has not changed: they now get 10 per cent more for their output, and they have to pay 10 per cent more for their input, namely labour. Therefore, they have no incentive to increase employment and expand output. So aggregate supply remains unchanged. And those workers who were jobless before remain unemployed.

This situation is described by a vertical aggregate supply curve in Fig.

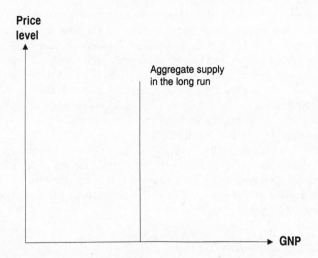

FIG. 5.3. The modified aggregate supply curve in a market economy

Note: If the general price level rises and wages increase proportionately, the real wage cost of firms remains unchanged. Firms have no incentive to employ more workers and expand production. Therefore, the aggregate supply curve is vertical in this case.

5.3. This figure resembles Fig. 3.3 in Chapter 3. An important difference should be noted, however. Figure 3.3 was used to describe a fixed supply of shoes in a planned economy, the supply being fixed by the central planning authority independently of price. Figure 5.3, on the other hand, describes a constant aggregate supply of goods and services in the economy as a whole. In this figure, the constancy of aggregate supply is the consequence of the behaviour of workers and firms who act in their own best interests and, as a rule, without government interference.

Which of the two stories told above is a better description of the supply side in the real world? The one where the aggregate supply curve slopes up (Fig. 5.2), or the one where it is vertical (Fig. 5.3)? This question, as we have indicated, cannot be answered unambiguously once and for all. Sometimes the first story seems to fit the facts reasonably well, sometimes the second. In any particular situation, the answer to the question depends on several things. Most important, perhaps, is the situation in the labour market.

In times of unemployment workers may be reluctant to insist on a wage increase when the price level rises. The rationale for such restraint on the part of the workers is a concern for the unemployed, who will remain out of work unless real wage costs are permitted to fall, other things being equal. Also, the reaction of workers or their union representatives in these circumstances will depend, among other things, on the influence of unemployed workers on wage settlements between workers and firms or between labour unions and employers' associations.

If full employment prevails, on the other hand, workers will in general see no reason why they should not demand full compensation for higher prices. According to this argument, Fig. 5.2 can be used to describe the supply side of an economy with unemployment, while Fig. 5.3 describes that economy in the case of full employment.

If joblessness is generally a temporary, or short-run, phenomenon, and if full employment is ultimately restored in the long run, Fig. 5.2 can be said to provide a description of the supply side in the short run, while Fig. 5.3 can be said to cover the long run. We shall make this assumption for the remainder of this chapter.[1]

5.4. Macroeconomic Equilibrium

We now have in place the two key ingredients necessary for discussing the market for goods and services: aggregate demand and aggregate supply. The reader can probably figure out much of the rest of the story. The next step is to merge aggregate demand and supply to show how the general price level and gross national product are determined in a free-market economy. This will pave the way for a discussion of two of the most important problems facing the general public and governments of the market economies today: inflation and unemployment.

The interaction of aggregate demand and supply is described in Fig. 5.4, where Fig. 5.1 has been superimposed on Fig. 5.2. Figure 5.4 looks identical to Fig. 4.1 in Chapter 4. The only difference is, once again, that we are now about to study the interplay between the *aggregate* demand and supply of goods and services in general, not just demand and supply in the market for shoes or shirts. Apart from that, however, Fig. 5.4 can be interpreted in much the same way as Fig. 4.1.

Let us begin by looking at the point of intersection of the two curves in Fig. 5.4. At this point firms employ as much labour as they want to, given the real cost of labour, which depends on the going wage and the prevailing price level. Firms therefore have no desire to move away from point A. Having agreed with their employers on wages and other working con-

---

[1] However, even in situations of substantial unemployment, workers may conceivably try to squeeze the highest possible wages out of their employers; compare the discussion of unemployment in Chapter 3. If workers do not believe that wage restraint would encourage firms to expand employment, or if the interests of the unemployed are under-represented in the wage-bargaining process, wages may increase in tandem with prices. If so, Fig. 5.3 is the correct one always.

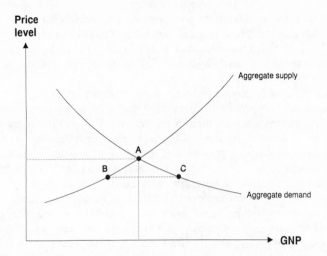

FIG. 5.4. Aggregate demand and supply in a market economy

Note: At point A the market for goods and services is cleared. The economic system is in equilibrium, in the sense that there is no tendency for the system to move away from point A.

ditions, workers have no incentive either to move away from the intersection point. Not all workers may be happy with this, however, especially those who are unable to find jobs at the going wage; but there is not much they can do about it.

As consumers of goods and services, workers are also content with point A: they are able to buy just as much as they are willing to at the prevailing price level. The same applies to firms and foreigners: at point A they too spend just as much as they want on domestically produced goods and services. Thus, they have no incentive to move away from the aggregate demand curve. In this sense point A is an equilibrium point. The intersection of the aggregate demand curve and the aggregate supply curve at this point determines the general price level and GNP simultaneously.

To see this more clearly, let us suppose that the economy happens to be out of equilibrium for some reason, say, at point B in Fig. 5.4. What happens then? We see that point B is on the aggregate supply curve. This means that firms are producing as much as they want at the price level prevailing at point B. Notice that the firms now produce less than at point A because they receive a lower price for their goods and services. Therefore, fewer workers have jobs at point B than at point A; unemployment is higher.

We also see that point B is *below* the aggregate demand curve. This means that households, firms, the government, and foreigners are buying less goods and services at point B than they would like to at the prevailing price level. At this price level, they would like to be at point C. Therefore, aggregate demand exceeds aggregate supply by the amount shown by the distance BC in the figure.

How does the economic system react to an excess of aggregate demand over aggregate supply? Producers realize that they can raise prices without losing customers; after all, there is demand for more than they are willing to supply at point B. So they raise prices. The initiative does not have to come from producers, however. Consumers may offer to pay higher prices in an attempt to get more goods and services. As the general price level rises, the excess of aggregate demand over aggregate supply levels off, both because supply increases and because demand decreases.

The upward pressure on the price level continues as long as aggregate demand exceeds aggregate supply. When the gap has been closed, the price level stops rising. Equilibrium has been achieved. This process is automatic, just like the microeconomic adjustment process described in Chapter 4. The economy gravitates to the intersection point A by itself, driven by the desire of firms and households to increase their material welfare.

From this we can conclude that, whenever the economy happens to be *below* the intersection point A in Fig. 5.4, it will move towards that point by itself, without outside interference. Readers should be able to convince themselves that the same applies when the economy is *above* point A: then, too, the system will gravitate to the point of intersection of the two curves in the figure. The 'invisible hand' that guides individual markets to equilibrium, as we saw in Chapter 4, also steers the economic system as a whole to an equilibrium where aggregate supply and aggregate demand are equal. So the economy reaches equilibrium by itself—at least, as long as there are no external forces at work to prevent this outcome.

5.5. Four Properties of Macroeconomic Equilibrium

The macroeconomic equilibrium achieved through the interaction of aggregate demand and aggregate supply is like the tip of an iceberg: there is more to the story than meets the eye.

A brief discussion of four interesting properties is offered here. First, we have seen how the general price level rises in response to excess aggregate demand. This is what is meant by inflation. Inflation is always and everywhere a consequence of excess aggregate demand. Without excess aggregate demand, there is no reason for prices to rise. Excess aggregate demand, in its turn, can stem from many sources, including government budget deficits and monetary expansion, as we shall see in Chapters 11 and 12.

Second, and very important, there is no guarantee that macroeconomic equilibrium is accompanied by full employment. It is quite conceivable that the economy may be locked into an equilibrium position where some workers are unable to find jobs for an extended period of time; compare Fig. 3.5 in Chapter 3. Persistent unemployment during the Great Depression in the 1930s or in Western Europe since the early 1980s is a case in point.

Third, the economy we are dealing with trades with the rest of the world. Domestic products are exported abroad in exchange for imports of foreign goods and services. In the equilibrium between aggregate demand and supply, imports can exceed exports, in which case the residents of our economy are accumulating foreign debt. Or the situation can be the other way around, in which case they build up foreign assets. Deficits and surpluses in the balance of payments *vis-à-vis* the rest of the world impinge on aggregate demand and supply, but there is no need for us to go into that now. The main point here is that there is foreign trade taking place behind the intersection of the two curves in Fig. 5.4.

Finally, we have also mentioned that the government levies taxes on its citizens. In return, the residents are provided with various goods and services—defence, education, health care, and so forth. But we have not said whether the government spends more or less than it receives in taxes in macroeconomic equilibrium. It could be either way. If government spending exceeds tax receipts, the deficit must be financed either by borrowing at home or abroad or by printing money. This has implications for aggregate demand and supply, also to be discussed in Chapters 11 and 12.

SUPPLEMENT 5.3. IS GNP A RELIABLE INDICATOR OF ECONOMIC PERFORMANCE?

The concept and measurement of GNP can be used for many purposes. When, in keeping with common international practice, we say that a country's GNP is so many US dollars per capita, we are referring to a statistical measure of the variable that appears on the horizontal axes of the diagrams in this chapter, divided by the population of the country. We must also be sure we know what year our US dollar figures refer to, because the value of the dollar (and other currencies) changes from year to year. A dollar today is worth much less than it was ten years ago because inflation has reduced its value during the period.

If used with sufficient care, measures of GNP per capita in different countries can be used to indicate the difference between the purchasing power of the incomes generated in the countries, and hence to compare the material living standards of the countries' inhabitants. This can be a tricky task, however. First, the measurement of GNP is based on market transactions. In an economy where households to a considerable extent directly provide themselves with goods and services, measured GNP will underestimate the material well-being of the population. Consider housewives who join the paid labour force: instead of growing their own vegetables in the kitchen garden, they now buy them in the market. Thus, GNP increases even if production remains the same.

Second, conventional measures of GNP do not reflect the damage done by pollution to the environment, or the depletion of natural resources (forests, fisheries, oil wells, etc.). For example, a nation can earn high incomes for a while by running down its resources, but then the country's future income potential is reduced. And the country's GNP per head, as recorded in official statistics, will be overstated in the sense that it cannot be sustained.

To take still another example, a country's GNP per head can be high compared with others mainly because its inhabitants work harder. In this case, a comparison of GNP per hour of work may provide a better indication of the material standard of living in the two countries than a comparison of GNP per capita.

GNP is not an all-encompassing measure of economic performance. The performance of two economies with identical GNPs per head may be different in other respects: one may be ravaged by inflation while the other has a stable price level; one may be accumulating debts abroad while the other is building up assets; one may have an unequal distribution of income and wealth compared with the other. And, finally, the inhabitants of one country may spend their hard-earned national income wisely while the people of the other, usually through their government, waste their income on things of limited use (excessive military spending, for instance). The last point, in particular, indicates why official figures on GNP per capita are generally not a satisfactory indicator of material living standards in centrally planned economies.

5.6. Effects of External Shocks

The above description of the interplay of aggregate demand and aggregate supply in a market economy should make it clear that *changes* in aggregate demand or supply will affect the price level and GNP. Imagine, for example, that the economy is initially in equilibrium and that aggregate demand increases. This can occur for various reasons. Foreigners may decide that they want to buy more of our products, perhaps because their economies are in an upswing. Domestic firms may decide to invest more than before in response to newly discovered profit opportunities. The government may decide to spend more money—to repair buildings and bridges, raise the salaries of teachers in public schools, or send men to Mars—or it may decide to lower taxes and so encourage households and firms to spend more.

Whatever the source of the increase in aggregate demand, its effects on the price level and GNP are described in Fig. 5.5. After the increase has taken place, the original intersection point A is no longer on the aggregate demand curve. At the original price level, aggregate demand now exceeds aggregate supply (which has not changed, by assumption). The excess demand, indicated by the distance AC, drives the price level up until a new equilibrium is reached at point B. So we see how an increase in aggregate demand increases the price level—that is, creates inflation.

As we have drawn the figure, GNP will also rise and unemployment will fall, provided there was some unemployment to begin with to make it possible for firms to find workers and so to be able to expand production. In this case, inflation is accompanied by a decrease in unemployment.

If, on the other hand, there is full employment at the outset, Fig. 5.6 provides a better description of what is going on. In this figure we have combined Figs. 5.1 and 5.3 which shows a vertical aggregate supply curve. In Fig. 5.6 an increase in aggregate demand, whatever its source, will raise

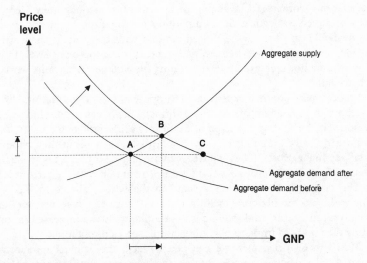

Fig. 5.5. Aggregate demand increases

Note: When aggregate demand increases, the price level rises, and GNP also rises, provided that idle workers are available.

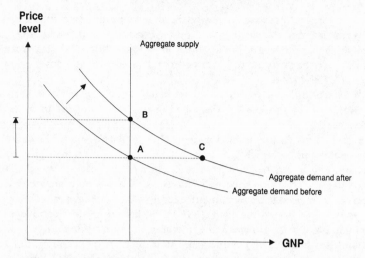

FIG. 5.6. Aggregate demand increases at full employment

Note: When aggregate demand increases, the price level rises. GNP remains unchanged because firms cannot find workers and therefore cannot expand production. The wage rate increases proportionally with the price level.

the price level as before. But GNP and employment are left unchanged because no workers are available to increase production—they were all fully occupied to begin with. Competition for the given number of workers available raises the wage rate in tandem with the price level. In sum, an increase in aggregate demand is bound to raise the price level, but it may or may not increase GNP depending on whether or not idle workers are available.

Let us make another experiment. Suppose the economy is initially in equilibrium and aggregate supply contracts. This can occur for several reasons. If, for example, the price of oil in world markets increases, as it did during 1973–4 and again 1979–81, firms that use oil in their production face higher costs. They will react by raising their prices and reducing their use of oil, and probably also labour, to contain costs. Therefore, the aggregate supply curve moves up and to the left, as shown in Fig. 5.7. The aggregate supply curve will, in fact, shift this way in response to any event that increases production costs—a unilateral wage increase that outstrips the ability of firms to pay, for example, or a crop failure.

After the oil price increase has taken place, the original intersection point A in the figure is no longer on the aggregate supply curve. At the original price level aggregate supply (at point C) now falls short of aggregate demand (at point A), which has not changed, by assumption. The

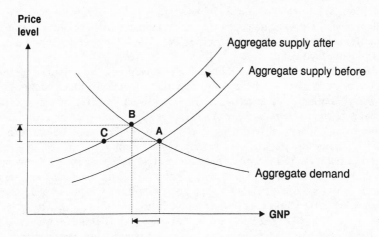

Fig. 5.7. Aggregate supply decreases

Note: When the cost of production increases, firms raise prices and contract production. Therefore, the price level rises and GNP falls.

ensuing excess demand, represented by the distance AC, drives the price level up until a new equilibrium is reached at point B.

Here we see again how excess aggregate demand increases the price level, and thus creates inflation. In this case, however, excess demand is generated by a contraction of aggregate supply rather than a boost to aggregate demand. Notice also that GNP is lower at the new equilibrium point B than it was initially, and unemployment is higher. In this case inflation is accompanied by an *increase* in unemployment.

In practice, aggregate demand and aggregate supply are always on the move. The equilibrium position of the economy is therefore in a constant state of flux. As a rule, both aggregate supply and aggregate demand increase over time so that GNP also increases. This is what is meant by *economic growth*. The increase of aggregate supply and demand over time tends to be uneven, sometimes rapid, sometimes slow, thus generating a cyclical pattern in the development of GNP through time. This is what is meant by *business cycles*. Usually, aggregate demand increases more rapidly than aggregate supply so that the price level also rises. The result is inflation.

5.7. Effects of Economic Policies

Now that we have seen how changes in aggregate demand and aggregate supply lead to movements in GNP and the price level, we are in a position to study the effects of economic policy on these variables. Clearly, if the government can influence aggregate demand or aggregate supply, it can also influence the price level and GNP and hence inflation and unemployment. To see this, take a second look at Fig. 5.5. Suppose the original equilibrium point A is characterized by high unemployment which shows no signs of abating. The government knows, as you do, that an increase in aggregate demand, regardless of its source, will stimulate the economy, thus raising GNP and reducing unemployment. But households, firms, and foreigners are buying just as much as they want at the prevailing price level at point A, and therefore have no interest in buying more. Knowing this, the government may choose to boost the demand for goods and services. It can do this directly by spending more money itself, or indirectly by lowering taxes in order to encourage households and firms to spend more. In either case, the economy moves from A to B in the figure.

It is not obvious, however, that the government would consider any of this the right thing to do. One reason why it may hesitate is that the price level will rise as the economy moves from A to B in the figure, and the government may consider the ensuing inflation worse than the unemployment.

Another reason is that increased public or private expenditure will lead to an increase in imports from abroad, and thus will weaken the country's trade balance or aggravate its foreign debt position, other things being equal. Yet another reason is that the government may fear that its intervention in the economy may be undesirable in itself, especially if it causes the public sector to expand little by little at the private sector's expense. Finally, the government may be convinced that the unemployment will ultimately disappear by itself in any case, so that no government action is necessary. If so, a government-induced stimulus to aggregate demand could make things worse by causing inflation after the unemployment problem had taken care of itself. Considering the alternatives, the government may choose to tolerate the unemployment.

This story can also be told in reverse. If inflation is a problem, the government can reduce it by restraining aggregate demand—by reducing public spending, raising taxes, or reducing the money supply (or its rate of growth). But this is generally not an easy thing to do, because a reduction of aggregate demand entails a contraction of GNP and a corresponding increase in unemployment, in the short run at least. This is why inflation is so persistent in many countries once it has taken root: governments often fear the consequences of stamping it out. Knowing that the increase in unemployment resulting from anti-inflation policies will be temporary may not be of much help if the government fears it will lose an election in the mean time on account of these policies. Besides, one cannot be sure that full employment will always and everywhere be restored automatically in the end. As we have mentioned before, economic history bears witness to many episodes of prolonged unemployment in market economies. In such situations, the government must choose between inflation and unemployment.

But would the world not be a better place if such painful choices could be avoided? Consider Fig. 5.8. Here the economy is in equilibrium with persistent unemployment at point A, initially. If the government could somehow increase aggregate supply rather than aggregate demand, and thus move the economy from A to B, both unemployment and the price level would fall.

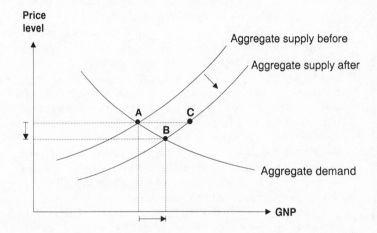

FIG. 5.8. Aggregate supply increases

Note: When the cost of production decreases, firms can lower prices and expand production at the same time. Therefore, the price level falls, as does the level of unemployment, since GNP rises.

In practice, however, finding effective ways to accomplish this feat has not proved easy. If it were easy, unemployment would not be as high as it is today in many countries in Western Europe and elsewhere. None the less, there are some things that governments can do to reduce production costs in the private sector and thus stimulate the economy from the supply side, such as lowering payroll taxes on employment. More importantly, perhaps, governments could try to encourage institutional and structural reforms to increase labour mobility between occupations and places. Also, reducing wage rigidities that lead to the exclusion of unemployed workers from the labour market would seem appropriate.

However, another potentially important method of stimulating the aggregate supply side of the economy is about to be put to the test in the European Community; see Supplement 5.4.

SUPPLEMENT 5.4. EUROPE 1992: SUPPLY MANAGEMENT IN ACTION

The European Community is making rapid progress toward its goal of establishing a single market by the end of 1992. By that time goods, services, capital, and labour will flow freely within the twelve countries of the Community. Their national borders will thus be effectively abolished.

The major benefit expected from freer trade within the Community stems from greater competition and greater efficiency which, in turn, will raise general living standards in Europe. This is expected to come about in two main ways. First, each Community member will have access to less expensive goods, services, labour, and capital in a single European market than would be the case if each country were confined to its own domestic market. Telephone service, for example, is 50 % more expensive now in some places than it is in the Community on average. Countries with expensive telephone services can save money by purchasing these services elsewhere in the Community where they are produced more efficiently.

Second, for some goods and services, production can be organized more efficiently, and less expensively, on a large scale in the Community as a whole than is possible in individual member countries.

Thus, through a more efficient international division of labour and through economies of large-scale production, it is envisaged that the general price level will fall and that aggregate demand and GNP will rise in Europe after 1992 (compare Fig. 5.8). By how much? This is difficult to say with certainty. One attempt to assess the magnitude of these effects indicates that consumer prices will gradually fall by about 6 per cent, other things being equal, and that the combined GNP of the Community countries will rise by about 4.5 per cent as a result, over a period of about five or six years. These effects could actually prove greater in practice if the establishment of a single market turns out to raise the rate of growth of the European economy because the effects of increased growth are cumulative.

Source: After Checcini Report (1988)

6

The Workings of a Planned Economy

I have always considered myself to be communist. That is my tragedy.

Edward Shevardnadze

ON the drawing-board a planned economy may seem to function well.[1] Indeed, as was pointed out in Chapter 4, one can imagine the 'invisible hand' replaced by an 'artificial hand', so that all the good properties linked to the effectiveness of a market economy are preserved. The goods produced can be distributed thereafter—in principle, to each according to his or her need.

6.1. A Lack of Incentives

The fundamental criterion the individual enterprise must relate to in a planned economy is the *quantity produced*, i.e. a target figure for physical production. Various bonus schemes will be linked to the achievement of this target figure.

Campaigns for quality have been introduced at regular intervals in Central and Eastern Europe. But since production has tended to show a steady decline in quality, the effect of such campaigns over a period of time must have been slight. As long as the target variable for the undertaking is the number of units produced, experience has shown that complicated or non-measurable aspects such as quality will be difficult to take into account.

Determination of the production target takes place by means of what must be called *negotiations* between the management of the enterprise and the centrally located planners. The enterprise will want the lowest possible figures: then the possibility of achieving them will be greater, and both managers and workers will gain bonuses. The planners, being pressured by ambitious politicians, often end up by setting unrealistically high production targets for the enterprises.

[1] In this chapter—as elsewhere in this book, unless otherwise obvious—we consider planning in simple and general terms. For a detailed account of the planning procedure in the former USSR, see Gregory (1990).

The result in many Central and Eastern European enterprises has been a downward adjustment of production plans towards the end of the year. Sometimes this will have a negative effect on bonuses, which may constitute 20–30 per cent of total wages. At other times the planning figures will be changed so that bonuses are paid out on the basis of the new, lower figures for total physical production.

Nevertheless, the enterprise will survive, no matter how well or poorly it does. This gives maximum job security. On the other hand, the certainty that, if necessary, the authorities will intervene and save the enterprise creates a form of passivity. The knowledge that funds will be injected if the situation should require it results in enterprises working under a 'soft budget constraint'.

When the target figure has been determined and divided into quarters and months, the enterprise will have little incentive to exceed these figures significantly. True enough, overfulfilment of the plan will give a solid bonus for the period in which the production realized greatly exceeds the target. But in the next round this will lead to higher planning figures—production previously achieved is, after all, the point of departure for the plan for the next period.

With few incentives for a significant increase in production, interest in inventiveness and innovation will also be dampened. Changes in the customary production process will always involve an element of risk. If the changes are successful and productivity shows solid progress, there will be a welcome one-time bonus, but this will be followed by correspondingly higher planning figures for future periods. If attempts at innovation are unsuccessful and planning figures are not achieved, bonuses will not be forthcoming and the management of the enterprise may have trouble with the planners. However, there will be no question of closure.

6.2. Uncertainty

In a centrally planned or command economy the individual enterprise works under a special form of uncertainty: namely, the uncertainty of whether the necessary inputs will arrive at the proper time, at the proper place, in a sufficient quantity, and with the expected quality. When all production is subject to a single, overall plan, this means an *absence of trade between enterprises*. There is, quite simply, no opportunity for enterprises to buy something on a free basis from other enterprises. In purely formal terms, the settlement of transactions between enterprises take place via bookkeeping over accounts in the unified banking system, at prices determined centrally.

The general—and particularly important—point here is that the enterprises lack alternative suppliers of the necessary inputs. If one link in the chain of production or in the distribution apparatus does not manage to achieve its targets, this will be transmitted through the system. Enterprises further down the chain of production will not receive supplies of necessary raw materials and semi-manufactured products, and their capacity to achieve *their* production targets will be correspondingly reduced.

For the individual enterprise, this form of uncertainty leads to the adoption of various measures, all of which serve to hinder the effectiveness of the system viewed as a whole. Let us describe two such measures. First, the enterprise will be encouraged to accept all the inputs it can lay its hands on. For the manufacturer of machinery, it will be a good thing to have an extra store of ball-bearings, screws, and nuts; then, if supplies of such items should fail to materialize for a period, there will be 'hidden reserves' to draw on. The producer of tinned goods will wish to have an extra store of raw materials to fill his tins with if supplies of fruit, meat, fish, or whatever should fail to arrive.

It is particularly easy in the latter case to see that such resources can easily come to be wasted. If supplies of raw materials go according to plan, the tinned goods manufacturer may easily end up with a store of spoiled surplus goods. It might have been possible for him to have produced additional tinned goods from this surplus store—but a lack of tinned cans might put a stop to that possibility. The machinery manufacturer's extra store will also involve a cost to society: too many resources are tied up in unproductive storage and may never be utilized. Instead, they may rust away or be rendered unusable in some other way.

SUPPLEMENT 6.1. THE PLANNED ECONOMY REQUIRES TOO LARGE INVENTORIES

Whereas total inventories in Soviet enterprises constitute more than 80 per cent of total production in the economy, the corresponding figure for the United States is 30 per cent. In the retail trade the proportion is the same: American shops have, on average, stores for about 30 days' sales, whereas the corresponding figure for the USSR is 92 days. To top it all, it is in the former Soviet Union that there is a scarcity of goods, both in production enterprises and in shops!

This is due, as we have explained in the text, to the fact that the planning system creates an uncertainty concerning future supplies, which leads to a build-up of stocks in enterprises. And if one important production component is lacking, it does not help to have overfilled inventories of other things.

The situation has deteriorated in the past twenty years. In this period an average of 6 per cent of national income each year has gone to the build-up of inventories in the Soviet Union, compared with 1 per cent in the United States.

If Soviet undertakings could come down to American standards as far as storage is concerned, a one-time release of resources corresponding to half the national income would be realized.

In the area of efficient storage, the Japanese have a good lead. At Toyota there are inventories of some components for only one hour's production ('just-in-time' production)! An advanced computer-controlled system which is based on continual production in small series has made this possible. Compared with American car manufacturers, the Japanese have only a tenth of stores of such components for a corresponding number of cars produced.

Source: After Shmelev and Popov (1989). This book contains a large number of other examples of how the Soviet planned economy systematically wasted resources.

The second measure that enterprises can adopt to safeguard themselves against deficiencies in supplies is to achieve a high degree of self-sufficiency. The machinery manufacturer finds it useful to have a small department producing various types of tools for the enterprise's own use, and the tinned goods producer likewise regarding tinned cans. The division of labour and specialization which planning aims at promoting will then be undermined by an exaggerated tendency towards self-sufficiency in the individual enterprise. Adam Smith's point concerning specialization and division of labour will scarcely be exploited. A concrete example may illustrate this:

Machine-building enterprises are overgrown with subdivisions that are not part of their main product. They are usually poorly mechanized subdivisions that produce, often by hand labour, instruments, fittings, castings, forgings, containers, and so forth. They are inefficient but they can produce almost anything . . . so there is no need to turn to trading partners. For example, in Moscow Province, half of all the machine-building enterprises produce their own instruments, forgings, and castings at a prime cost that is 1.5–2.5 times higher than in specialized enterprises. (Shmelev and Popov 1989: 118)

Since uncertainty concerning supplies can soon lead to different enterprises having different extra stores of different products, the situation is ripe for *barter* between enterprises. The tinned goods producer has a surplus of aluminium sheets for the production of tinned cans. The machinery manufacturer could use these aluminium sheets. But what can he offer in return? Well, extra production of reserve parts for tractors, which the collective farm needs, could be utilized in return for fruit production in excess of the planning target for the farm. And the tinned goods factory is interested in this fruit production. Thus, the way will be paved for a three-way barter. Without money as a means of exchange, the trading opportunities of firms are thus severely restricted.

6.3. The Consequences of Too Low Prices

In planned economies, where money and prices are not allowed to play a part in trade between enterprises, special middlemen see it as their job to arrange such forms of barter. They often work on the border of the permissible, or beyond this border. But, since the effectiveness of the system viewed as a whole is served by such unofficial 'huckstering', the authorities will often turn a blind eye to it. In addition to the fact that such activities underscore the system's own shortcomings, it also paves the way for various forms of corruption.

SUPPLEMENT 6.2. A HIDDEN ECONOMY IN BOTH THE EAST AND THE WEST

Corruption and extensive activity in the 'hidden economy' are not unknown in countries with market economies, either. Calculations from various countries indicate that such activity lies in the area of from 2–3 % of national income to 20–25 %. The 'hidden sector' in countries based on planned economies probably lies in the upper part of this interval, or perhaps even higher.

However, the reasons for the growth of the hidden economy are different in the two systems. In a planned economy, hidden and corrupt economic activity will be due mainly to the lack of availability of the goods and services in question on the open market. In a market economy, all lawful goods and services are largely freely available, but direct and indirect taxes provide an economic incentive to hide transactions. When the authorities impose both a value added tax (in many countries of 20%) and an income tax of 20–40% (in many countries of more than 50% on the highest band of earnings), there will be huge sums to 'save' by keeping transactions hidden.

For consumers, too, the absence of alternative suppliers is a problem. Shops, like producers, receive the goods specified in the plan, and sell them at prices which the plan also has stipulated. As far as many goods are concerned, the belief that large enterprises are more effective than small ones has led to monopoly-like situations in the economy. This means that only one product or just a few alternatives are available for consumers to choose from, and all are produced by the same firm.

When the supplier of, say, TV sets has such a monopoly, it will not help much for the individual shop to complain: the shop does not have any other producers to contact. The shop and its customers will be at the mercy of the monopolist.

SUPPLEMENT 6.3. EXTENSIVE BARTER CHARACTERIZES A PLANNED ECONOMY

In the spy thriller *Russia House* by John le Carré, we find a telling description of how consumers must engage in a long series of barter transactions in order to end up with the goods they really want.

> She had taken her decision. As soon as she got to the office she would collect the two tickets for the Philharmonic, which the editor Berzin had promised her . . . At lunchtime after shopping she would trade the tickets with the porter Morozov who had pledged her twenty-four bars of imported soap wrapped in decorative paper. With the fancy soap she would buy the bolt of green check cloth of pure wool that the manager of the clothing shop was keeping locked in his storeroom for her. Katya resolutely refused to wonder why. This afternoon after the Hungarian reception she would hand the cloth to Olga Stanislavsky who, in return for favours to be negotiated, would make two cowboy shirts on the East German sewing machine she had recently traded for her ancient family Singer, one for each twin in time for their birthday. And there might be enough cloth left over to squeeze them both a private check-up from the dentist.

In recent years control of wage payments measured in relation to physical production has been lost in the ex-Soviet Union, among other places; wages have risen much faster than production. When prices are kept artificially low at the same time, this means longer queues. The combination of relative scarcity of goods and relative abundance of money has caused payment in roubles to become a steadily smaller part of the sacrifice that is required to gain access to goods: queuing-time constitutes an increasing part of 'payment' for consumers.

Another consequence of the fact that prices are usually lower than supply and demand would indicate is that people buy whatever they can get anyway, even though they actually want something else. This results in a form of *forced consumption*. Or else people leave the shops without obtaining what they want and have nothing else to do with their roubles than to deposit them in the bank or consume more services, for example by going to cultural events or sports competitions. Some people invest in art or precious stones in the hope that such investments will take better care of the purchasing power of their savings over a period of time.

When prices are kept artificially low, those who stand behind the counter acquire considerable power. They can set aside some goods for which there is a great demand, then sell them to friends and relatives. Or 'special customers' may be offered these goods after closing-time, perhaps at a price higher than the official one. The employees in the shop can put any extra profits into their own pockets.

There is an important general point here: when prices are kept low so that the market is not cleared, power is redistributed. In a rationed economy, the employee behind the counter gains power over the customer standing in front of the counter; and the bureaucrat and planner gain power over the enterprises through the determination of the quantity of physical supplies to be distributed from enterprise to enterprise. In a market economy, on the other hand, where all the producers are fighting to win customers, such personal power is replaced by competition between producers; and the possibility of corruption arising from scarcity of goods and centralized supplies becomes correspondingly smaller. In a market characterized by competition, power lies with those who stand in front of the counter—the customers.

For the efforts made at the workplace, too, the scarcity of consumer goods has significance. The incentive to work hard and gain a bonus is dampened by the knowledge that these extra roubles cannot easily be traded for the desired consumer goods. (In economic terminology, we say that the roubles are not *internally convertible*.) The shopping trip is time-consuming and the chances of finding what you are actually looking for are so small that the extra effort in your job that would be necessary to gain a bonus may not be worth the trouble.

Here we can see the contours of a vicious circle; virtually empty shelves in the shops reduce motivation at the workplace, which again ensures that there will be spaces vacant on the shelves. The idea that prices should be set free only after the queues are gone turns the problem upside-down; as long as prices are determined centrally and are kept lower than the prices that would result in a clearing of the markets, the queues will always be there. However, setting prices free at one fell swoop has not been regarded as a solution either—except in Poland as of 1 January 1990 and in Russia two years later—out of fear of the reactions this would cause.

6.4. Gigantism

Up to the beginning of the 1960s, the centrally planned economies could congratulate themselves on substantial growth. Large enterprises based on relatively basic technology which produced a fairly limited range of simple goods were the solution of the times. Since the consumers' preferences at this stage were quite stable, the advantages of large-scale operation could be exploited; that is to say, costs per unit produced fell when the number of units produced increased.

However, there is a limit to this type of large-scale production of relatively unsophisticated goods. This means that up to a certain point it does pay to increase production, but after this point unit costs will rise with further production. That point defines the optimal size of an enterprise. This is something that is difficult to determine. Moreover, the optimal size will vary from branch to branch and over time in the same branch.

In the former Soviet Union, an average of 1,000 persons work in an industrial enterprise, and 500–600 on a collective farm. Measured according to standards in market economies, these are quite unbelievably high figures. Since competition in a market economy applies pressure to keep costs down, it is hard to avoid the conclusion that there is a considerable lack of small and medium-sized enterprises in Central and Eastern Europe, among other things.

SUPPLEMENT 6.4. SMALL IS BEAUTIFUL

In countries with planned economies, the authorities have concentrated on large-scale operations to a far greater extent than managers of private undertakings in Western market economies have found profitable. A comparison of size distribution in the different countries has been carried out by the Hungarian economist Eva Erlich.

Her work shows that in the socialist countries of Czechoslovakia, East Germany, Hungary, and Poland, on average, half the labour force work in undertakings with more than 1,000 employees. The comparable average figure for Austria, Belgium, France, Italy, Japan, and Sweden is 19 per cent.

Further, in these market economies 68 per cent worked in enterprises with between 10 and 500 employees in 1970. In the socialist countries the figure was 34 per cent, or exactly half as much.

Another example of the difference in enterprise size appears if we compare state-owned industry in Poland with Western industrial enterprises. On average, there were approximately 380 employees in the Polish enterprises compared with approximately 80 in the Western ones throughout the 1980s. Compared with the market economy in South

Korea, where one-third of the industrial workers work in undertakings with less than 100 employees, the corresponding fraction for Poland was one-tenth.

What explains these differences between the planned and market economies? One factor is the merger after the Second World War of minor undertakings in Central and Eastern Europe to form large conglomerates. Another is the Stalin period's focus on heavy industry. In planned economies heavy industry constitutes a far greater proportion of total industry than in most market economies. Among other things, for technological reasons heavy industry is organized in large units. Experience gained from 'mini-steelworks' in Italy, however, would indicate that such undertakings can also be operated profitably on a smaller scale.

A third reason for relatively large undertakings in countries with planned economies is that large-scale operations are often considered synonymous with efficient operations. However, in reality the situation is often the opposite: efficiency falls when the whole society's production of a product is gathered into one or just a few large units. The reason is that the efficiency-promoting effect of competition between many enterprises is eliminated.

In Central and Eastern Europe the situation is made worse by the fact that imports have been prohibited. The availability of imported goods would give customers and consumers freedom of choice, thus exposing even the largest and most dominating producers to competition. Experience has shown that free trade between countries is the best anti-monopoly policy a government can apply.

Source: After Erlich (1985: 267–95)

What could be the reason for this gigantism? Let us examine three factors. First, centralized planning will undoubtedly be somewhat simpler—at any rate on paper—if one is to keep track of 1,000 enterprises each with 1,000 employees, rather than 10,000 enterprises with an average of 100 employees (not to mention 100,000 enterprises with 10 employees each). Among other things, since many small units are more difficult to handle in a planned economy than just a few large ones, the small family farms have had to give way to the large collective farms. However, the belief that large, state-owned units in agriculture would give returns in the form of increased productivity has not materialized. When the farmer loses the stimulus as owner of the land to reap its returns himself, productivity in agricultural production falls. This is an experience shared by all countries in which landless agricultural workers are responsible for the bulk of agricultural production.

Another factor that explains the gigantism in Central and Eastern Europe is the desire of enterprises for a maximum degree of self-sufficiency. This, combined with 'soft budget constraint', gives management hardly any limitation on the desired size of the firm.

Third, the one-sided concentration on volume in production has led to a discrimination against the service production that usually takes place in smaller undertakings. According to Marx's theory, service production does not contribute to value added. Thus, the service sector in planned economies has remained underdeveloped. In Central and Eastern Europe the service sector is normally responsible for about 20 per cent of the total value added: in many developing countries the size of this sector is 40 per cent; and in the Western market economies, it is going on 60 per cent. A well developed system of various services, such as transport, telecommunications and distribution, banking, and insurance, is of vital importance to the effective utilization of the total resources in the economy. Service industries ensure that manufacturing enterprises are given good conditions to work under.

6.5. Summary

The system of economic planning is characterized by three elements:

1. *targets for physical production* as the decisive criterion for fulfilment of the plan;
2. *absence of price* as a signal of relative scarcity, that is as a mechanism for clearing the markets, as discussed in Chapter 4;
3. *absence of bankruptcy* as an impersonal mechanism for the dismantling and reorganization of production: this is a consequence of 'soft budget constraint'.

These three factors will lead to a state of *chronic excess demand* with corresponding shortages of both inputs and finished goods. When, in addition, many branches consist of large enterprises with a monopoly-like position in the economy, the way is paved for a situation characterized by a *seller's market*. This means that at all stages producers' interests will prevail at the expense of consumers.

The planning system in principle should involve continuous feedback to the central authority, but it does not function like this in practice. Adjustments in production and in prices do not take place. When the plan is made for the coming year, the whole situation is frozen. This means that the individual enterprise is deprived of the possibility of choosing between alternatives, both as far as its own production is concerned, and as far as the availability of inputs is concerned. Rather than adjusting flexibly to changed conditions in technology, organization, and demand, the system takes on the rigidity of the plan itself.

The result of this system of organizing a country's production is that economic growth takes place through the utilization of more resources, and not through better utilization of existing resources. It has been shown that the alternative, *intensive growth*, which involves growth through increased productivity, cannot be achieved successfully by a planned economy. Again we draw on Shmelev and Popov (1989: 139–40):

In construction they don't build, they 'utilize resources'. Outlays are increasing, investments are growing, construction material factories cannot work fast enough to fill the demand for brick, cement, and reinforced concrete units. The scale of construction activity is expanding. . . . Only one thing is missing: a commensurate increase in the number of finished products. Resources disappear without a trace in the production process itself, in the 'black holes' of the construction complex.

This situation can be read from figures that compare energy and raw materials intensity in Central and Eastern European production with that of Western Europe. Roughly calculated, twice as much energy and raw materials are utilized per unit of finished product in the East as in the West. This indicates very inefficient resource utilization and consequently great pollution problems. In addition, the quality of finished products is poorer. To make matters worse, the difference in the capacity to utilize inputs and the difference in quality have both increased in the past couple of decades. It does not help much to have an impressive coal and steel production when these resources can only be refined into a relatively small amount of finished goods of poor quality.

In a planned economy, where prices play a less important role, the problem is that the various alternative forms of utilization of resources cannot be measured against one another. Is the value to society of an increase in shoe production of 100,000 pairs per year greater or less than the value of increasing production of shirts by 200,000 units per year? To be able to answer this question, we must have a knowledge of consumers' preferences or desires. But such knowledge is not available in an economy where prices are determined centrally, with the result that there is a shortage of both shirts and shoes. Even if the planners *were* capable of making the right decisions, they would still lack the large body of information necessary to do so.

In a market system, where economic calculations are made in terms of market prices and money, it is much easier to compare the usefulness of the alternative ways in which resources can be utilized. And when the economic agents themselves are striving to achieve the highest income possible, it turns out that the need for planners who can control the whole thing from the centre disappears. Part II is dedicated to a closer examination of the way in which a market economy works.

PART II

The Workings of a Market Economy

THE market economy has proved to be a flexible and robust way of organizing production in a modern society. An important reason for its flexibility lies in the bankruptcy mechanism: firms that are unable to compete successfully, i.e. that lose money, are forced to reorganize their operations, and ultimately to close down if all else fails. Labour and other resources utilized by the unsuccessful enterprise will thereby be released for new firms. The drawback is that closures will leave some people temporarily out of work.

However, bankruptcy does not have to entail closure of an undertaking. In many cases the firm can continue under new ownership and new management. If final closure is the result, however, those who lose their jobs will often be unemployed for a while before they find new jobs. Thus, the flexibility and robustness of the economic system is at the expense of the job security of the individual.

For a market economy to function, three things must be in place: private ownership, competition and free exchange of goods, and a 'hard' budget constraint which permits bankruptcies, and thus creates conditions for necessary economic renewal and growth.

The role of prices in a market economy can scarcely be overestimated. Through prices, both producers and consumers receive the information necessary for their own decisions. If enterprises are to survive in competition with others, they must produce goods and services that consumers will buy. Thus, consumers are the focus of attention in a market economy. In the words of Adam Smith, 'Consumption is the sole end and purpose of production; and the interest of the producer ought to be attended to only so far as it may be necessary for promoting that of the consumer.'

A successful transition from a planned to a market economy requires a vigorous *supply response* in the economy. This can only occur in an economic environment where new, private firms get started and expand where profitability seems to be good. It is conceivable, however, that the mentality of business managers and of the population in general, especially their passive or even hostile attitude towards entrepreneurship, initiative, and risk, reduces their capacity for supply response in the process of adjustment that is now taking place in Central and Eastern Europe. Specifically, there is a certain danger that decades of stifled private initiatives may have produced a widespread feeling of 'learned helplessness', to borrow a phrase from experimental psychologists.

We shall not conceal the fact that conflicts and dilemmas are inherent in a market economy. The most important ones will be discussed in Chapter 9. However, the market mechanism can often be harnessed to meet economic challenges to society. The authorities can create favourable conditions for adjustment and reorganization of private enterprises, for example by imposing a tax on production activities that cause pollution, while still preserving the freedom of the individual enterprise to make its own decisions with the aim of obtaining maximum profits.

In contrast to some of his successors, Adam Smith did not make a graven image out of greed. However, he argued forcefully that self-interest could be utilized to establish an appropriate way of organizing production in society. As the world has gained experience of the alternative—central planning—Adam Smith's arguments and insights have proven especially relevant and useful.

7

Prerequisites for a Market Economy

> You cannot have capitalism without capitalists.
>
> *The Economist*

FOR those who wish to ridicule the market economy, it will be tempting to have recourse to the expression *laissez-faire*, meaning that the authorities shall leave the arena entirely to the private sector. That is not the way things are, however. If an economy based on the free play of market forces is to function well, the authorities must have a strong presence in the arena, at least in order to solve the following three problems: private ownership must be guaranteed; competition and the free exchange of goods must be a reality; and a well-ordered code of regulations governing bankruptcy and liquidation of firms must be in place. Let us look more closely at each of these three prerequisites.

7.1. Private Ownership

Following our discussion of Table 4.1 (in Chapter 4), we concluded that, if prices, rather than centrally determined planning figures, are to do the job of co-ordinating the bulk of decisions taken in the economy, private ownership of the means of production will be necessary. In a society where, over a period of several decades, private ownership has been limited to personal goods and chattels and perhaps a small plot of land, understandably, there will be uncertainty and a lack of knowledge about what private ownership of the means of production actually involves. The task of the new authorities will be to install an appropriate code of legislation guaranteeing private ownership and instilling confidence in people that this right will not be curtailed or withdrawn from them at a later date. Firm and credible regulations governing this issue will be highly important.

Experience of reforms of planned economies has made the populations of many countries sceptical of the authorities. This is due to the fact that decisions which have given greater freedom to the individual have later been reversed. When the planners saw the productive forces that were released by an extension of private ownership, they felt threatened, and therefore saw to it that the regulations were reversed. In the Soviet Union, for example, the right to till one's own plot of land was first extended and then curtailed again. Today, the incentive for individual families to improve the land and make investments, say, in the form of irrigation facilities, so that over a period of time the land can provide better crops, will be substantially reduced if they cannot be sure that they will be able to till the same plot of land in years to come.

Nor will the person who sets up a small workshop for car repairs have any incentive to think far ahead when, with good reason, he fears that the authorities will take over the workshop if it goes too well—or that the authorities may place obstacles in the way, such as a surprising increase in taxes, a moratorium on supplies of necessary inputs, or a substantial increase in the price of such inputs. Rather than putting aside money accruing from the daily operation with a view to extending the undertaking, enterprising persons will choose to consume most of what they earn or to hold on to their wealth in the form of valuable objects. In a planned economy, there is quite simply no room for the accumulation of capital in private hands in the form of increased investments so as to provide a basis for increased production.

For the creation of a market economy, one must have private proprietors of both land and enterprises. And if the capital-owners are to play their part in creating a healthy and growing economy, they must be allowed to expand; that is, they must be given the chance to accumulate more and more capital. Unprofitable investments, on the other hand, must be allowed to result in owners losing a part or all of the capital they started with. However, growth cannot be completely unfettered: it is the duty of the authorities to see to it that tendencies towards the formation of monopolies are combated. This is because, with only one supplier of a product (a monopolist), customers will have no alternatives to choose from. This will give considerable power to the monopolist, and the economy will become producer-oriented. This in turn will lead to high prices, poor quality, and inefficient utilization of resources.

7.2. Effective Competition and Free Exchange of Goods

This brings us to the second area where the authorities must intervene in a market economy: namely, a code of regulations which will ensure effective competition and free exchange of goods. In the absence of such a code, producers in an industry will often be able to increase their profits by reaching agreements with one another on limiting competition.

An oft-quoted statement by Adam Smith will illustrate this point in an elegant manner:

People of the same trade seldom meet together, even for merriment and diversion, but the conversation ends in conspiracy against the public, or in some contrivance to raise prices.

Let us return to our example of shoe manufacturers from Chapter 4. Here we saw that the market will gradually arrive at a state of equilibrium where 20,000 pairs of shoes are produced each month at a price of 100 roubles a pair. Let us presume that there are eight shoe manufacturers in this market, and that one evening they meet for a confidential conversation. The following calculations are set up:

The present situation: Sale of 20,000 pairs of shoes at 100 roubles a pair will give a total income of 2,000,000 roubles per month.

The alternative: By selling 17,000 pairs of shoes, it would be possible to achieve 120 roubles a pair. This would give a total income of 2,040,000 roubles.

By reducing sales by 3,000 pairs of shoes, income will increase from 2,000,000 roubles to 2,040,000 roubles, i.e. by 40,000 roubles. Higher prices and reduced production would mean that the shoe manufacturers as a whole would augment their profits.

In practice, they could do this in several ways. For example, they could all agree to reduce their production by 15 per cent (3,000 is 15 per cent of 20,000); or a manufacturer or two could, in return for payment by the remainder, close down production; or one enterprise alone could buy up the other seven.

Although in this way producers would improve their own economic situation, consumers would be worse off. For society viewed as a whole, production of shoes would be less than desirable. And the price to consumers would be higher than under free competition.

This example illustrates an important property of a market economy: namely, that the individual businessman, who in principle is a supporter of free competition, will often seek to limit competition in his own market. Here the authorities have an important duty to see to it that this does not happen. For this reason, there is a need for laws which will prevent monopolies and competition-inhibiting collaboration both between enterprises, and between enterprises and the government.

In the production of certain goods, technology may indicate that extremely large-scale production is necessary if costs per unit produced are to be as low as possible. We then talk about *economies of scale*. To put it a little more technically, we have economies of scale when the costs per unit produced fall with increased production, and reach a minimum when there is a considerable volume of production. For example, this would seem to be the case in connection with the production of nuclear energy, aeroplanes, cars, computers, and so on.

In a small country one may arrive at a dilemma here. On the one hand, it would like to achieve the lowest possible cost per unit produced. This may indicate that there should be only one or just a few domestic producers of the product. In the case of Sweden and cars, for example, there is only room for two manufacturers: Volvo and Saab. But with only one or two domestic producers, there will be little competition. In other words, the way will be paved for monopolistic adjustments by the producers. This means that they may limit production and raise prices compared with what would happen in a market characterized by stiff competition.

However, the dilemma can be solved. By allowing *imports from abroad*, it is possible to have only one or two domestic producers without this leading to a monopoly or greatly reduced competition which would make goods more expensive for customers. Continuing with the example of Sweden and cars, if the situation was such that Swedes were only permitted to drive around in Swedish-produced cars, it is very possible that Volvos and Saabs would be expensive and perhaps of poor quality. As companies, Volvo and Saab might exploit their monopoly position and gain substantial profits.

For Sweden viewed as a whole, this would be a poor solution; too may resources would go to produce too few, substandard, and expensive cars. To top it all, Swedish exports of cars would scarcely have reached great heights—quite simply because experience has shown that a protected monopoly situation on the domestic market reduces international competitiveness.

However, through imports Swedes can buy Mercedes, Toyotas, Chevrolets, Renaults, or whatever else they wish. In other words, competition in the car market in Sweden is extremely stiff. This competition has meant that Volvo and Saab cannot rest on their laurels: they must keep working continuously to produce a better product at a lower price. This they have largely managed to do, which is clearly demonstrated by the considerable *exports* of Swedish cars to other countries.

SUPPLEMENT 7.1. SPECIALIZATION AND TRADE

Industrial firms in Western market economies often specialize strongly in just a few products. This provides a basis for international trade. In some branches production is aimed largely at exports. In others, domestic consumption is totally dependent on imports. Many industrial firms import a relatively large proportion of their inputs, either from their own subsidiaries abroad or from wholly independent foreign subcontractors.

Sweden has specialized strongly in, among other things, the production of paper for newspapers (newsprint). Of the total production, 86 per cent was exported in 1987. All domestic consumption of newsprint is satisfied through production by Swedish enterprises. On the other hand, production of clothing and footwear in Sweden today is very modest in relation to domestic consumption. (*Consumption* is defined as production plus imports minus exports.) In general, Swedes wear clothing and footwear produced in other countries. In simpler terms, it could be said that a part of the export revenue on newsprint is used to pay for imports of clothing and footwear from foreign manufacturers.

Cast iron is a classic industrial product. Although Sweden has a significant steel industry, domestic production of cast iron is extremely small. The cast iron that is needed, for example, for the sophisticated Swedish production and export of seamless steel pipes is cheaper to import.

Sweden has a substantial production of private cars (Volvo and Saab). However, in spite of the large domestic production, imports are also considerable. Why did Sweden import cars worth Skr14 billion in 1987, when domestic car production amounted to Skr24 billion? This is an example of intra-industry trade; that is to say, the same type of product is both exported and imported. For example, many Swedish consumers prefer German, Japanese, or French cars to Swedish ones; thus, these cars are imported to Sweden. At the same time, some German, British, and American consumers prefer Volvo and Saab to their national makes.

In order that there is good competition and free exchange of goods, the existence of correct and satisfactory information to consumers is important. Here, too, the authorities will have a role to play. Regulations are needed to protect consumers from false and misleading advertising. Volvo recently experienced an example of the necessity of such regulations in the United States. There, an advertising film was made showing that Volvo was more robust in collisions than other cars. This happens to be true. However, the Volvo that was used in the film had been specially constructed for the purpose. When this became known, Volvo apologized for the advertisement and withdrew the film immediately. And the advertising agency that had made the film for Volvo was fired. Another agency was engaged instead.

7.3. Bankruptcies and the Reorganization of Firms

The third area where the authorities must work actively to ensure that the market functions has to do with bankruptcies and the reorganization of undertakings. In a market economy it is normal for between 2 and 6 per cent of all firms to go bankrupt in the course of one year. (The vast majority are small and relatively newly started enterprises.) This is due to the fact that they are unable to honour their financial commitments and are forced for that reason to cease business.

The reason why bankruptcies take place so frequently in a market economy is connected with *uncertainty*. Whereas engineers and researchers in natural science and technical subjects can arrive at virtually exact laws and relationships, in the social sciences one has to make do with less precise knowledge. The engineer can calculate exactly the horsepower of a car, the efficiency of combustion, and the power of acceleration. The economist, on the contrary, possesses far from certain knowledge concerning how expensive the car will be to develop and manufacture, and whether it can be sold in sufficient quantities at the price that has been calculated.

An erroneous estimate of costs, and most of all of market prospects, which leads to sales problems, will mean that society's resources will have been utilized inefficiently. The enterprise will lose money and will have every reason to reorganize production.

It is only when the enterprise does not manage to carry out this reorganization, that is to say when it continues to lose money, that the bankruptcy mechanism will come into force. This means that the creditors, i.e. those who are owed money by the enterprise, will demand that the enterprise be liquidated. The legal system is brought into the matter by having a special court for bankruptcies, the probate court, take over management of the enterprise. The probate court, or the official receiver, considers how the creditors can get as much as possible for their claims. Sometimes the enterprise will be sold as one unit; at other times the property will be sold piecemeal. The money which thus comes in is divided among those who have had their claims on the enterprise approved. Should the total claims amount to 200,000 roubles, and the revenue from selling the enterprise's property only amounts to 40,000 roubles, each claimant will receive only 20 per cent of his claim. The money will quite simply go no further.

SUPPLEMENT 7.2. 'HARD' VERSUS 'SOFT' BUDGET CONSTRAINT

The distinction between 'hard' and 'soft' budget constraints, in the context of economic systems, is an important one.

'Hard' budget constraint means that enterprises cannot count on assistance from the authorities if their earning capacity fails. They will be left to themselves and must take the consequences of their financial and economic dispositions. This will result in enterprises, in the struggle to survive, being forced to change production when prices, technology, or other external conditions are changed.

If shoe manufacturers in a market economy begin to experience that more customers would rather have boots than sandals, they will be forced to change production in that direction. The customers' preferences will determine what is to be produced.

If their budget constraint is 'soft', on the other hand, enterprises are under much less pressure to adjust production to changes in external conditions. This is because if the enterprise runs into problems the authorities will be ready to intervene—by increasing the centrally stipulated prices, by awarding special loans or grants, or by renouncing demands for payment of taxes.

Should our shoe manufacturers be working under a 'soft' budget constraint and boots become more popular than sandals, they will therefore not have to worry so much about it: they know that losses arising from unsold sandals will be covered by the authorities anyway.

Over a period of time, the *mentality* of enterprise managers will be characterized by the type of budget threshold they face. A transition from a planned to a market economy will mean that 'soft' budget constraints will be replaced by 'hard' ones.

There are two disadvantages and one advantage of a bankruptcy. The disadvantages are that people may—but will not necessarily—lose their jobs, and that creditors will—quite certainly—get only part of their money back. The advantage is that production in unprofitable enterprises is terminated; thus, the resources that were being utilized in them will be released. In some cases new owners will take over the bankrupt enterprise and continue to operate it, but in a more effective manner. In other cases the firm will be closed down.

The advantage of closure of unprofitable undertakings is so important that it outweighs the two disadvantages entailed by the bankruptcy mechanism. In order to see this clearly, it will be useful to make a comparison with the situation in an economy where bankruptcies are not permitted to happen. In such an economy, characterized by 'soft' budget constraints, undertakings with poor production records will not be weeded out. Over a period of time, therefore, resources will be locked up in the production of goods and services that society has little use for.

As an illustration, let us again look at the shoe manufacturers from previous chapters. If, as a point of departure, there were eight enterprises, and after a while two of them lost control of income and expenditure, in a market economy they would go bankrupt. People who were owed money by these two enterprises would have to suffer a loss, and the workers would lose their jobs. Considering the knowledge and skills the workers have, some of them might be employed by one or more of the six remaining enterprises; production in these would probably increase when two rivals closed down. Others might have to go unemployed for longer, perhaps having to undergo a programme of retraining before they were back in work.

It is also possible that the person or persons who, during the bankruptcy case, bought the shoe factory itself, with machinery and equipment, would start up new production. In such a case many people would get a new *employer* but retain their old *jobs*. If new owners are to be more successful than the old ones, however, they must demonstrate greater proficiency in running the enterprise.

As for the loss that creditors will suffer, it is better for society that this loss be sustained than that unprofitable production should continue. The creditors will naturally be annoyed. But they will have learned a lesson: namely, to be careful about whom they lend money to. This will make them more cautious the next time. In this way the bankruptcy will have a useful educational effect. We could say that it will force the participants in the market to mutually monitor each other's undertakings so that the system as a whole will become more efficient in its utilization of resources.

However, it is impossible to free oneself from uncertainty in the economy. And lenders who have lost money on one client will risk losing money on another later on.

Bankruptcy is a necessary element in a market economy. This element ensures that there will be a purge and cleaning-out of inefficient firms. In the short term, and in each individual case, bankruptcy seems a brutal mechanism that strikes unjustly. Working men and women, in spite of strenuous efforts, are suddenly without their old jobs and their old employers. Lenders, among them other enterprises that have supplied inputs to the bankrupt enterprise, lose money. But since uncertainty is an unavoidable element in economic decision-making, it is not possible to avoid the fact that from time to time labour and capital will be invested in the production of goods and services that turn out to be unwanted.

7.4. Summary

The market does not emerge in a vacuum. In order to ensure real and effective competition, there is a need for carefully thought-out laws. Producers must not be given an opportunity to collaborate in limiting competition. Private ownership must be guaranteed. And the danger of going bankrupt must be real.

SUPPLEMENT 7.3. MANY NEW LAWS ARE NEEDED

In the text we have stressed three areas where the authorities must introduce suitable legislation that will regulate the market and ensure that it functions according to purpose (private ownership, competition, and a bankruptcy mechanism). However, the number of laws that will be necessary is much larger. In a paper by Anders Åslund, thirty or so laws are thought to be necessary. The most important ones are:

- a law on trade, emphasizing free competition
- a law on ownership, guaranteeing property rights
- anti-monopoly legislation
- a law on joint stock companies
- a law on co-operatives
- a law on private firms
- a law on land ownership
- a law on intellectual property rights and patents
- a law on the stock exchange
- a law on bookkeeping and auditing
- a law on privatization
- tax legislation
- a law on banking
- a law on bankruptcy
- a law on customs and foreign trade
- a law on foreign acquisition and ownership

Source: Åslund (1990)

The duty of the authorities to ensure the existence of a well functioning market will not be concluded as and when an appropriate legislative code is established. They must continually monitor adherence to the regulations. And at times there will be a need for changes in the legislative code. An example would be when countries enter into trade agreements with one another: then customs rates and other trade barriers must be reduced or removed altogether.

8

Challenges Posed by the Market

EVEN though the prerequisites for a market economy discussed in the preceding chapter are in place, goods and services will not present themselves to the population by magic. The people will have to produce them. In this chapter we shall examine some of the challenges facing countries and peoples when a market economy is to replace a planned economy.

Effective competition and free exchange of goods constitute the core of a market economy. When this core is in place, the economic decisions can be co-ordinated by having each participant do what serves his self-interest best. 'Economic Man' as a *consumer* strives towards the greatest possible utility, when income and prices are given; 'Economic Man' as a *producer* strives towards the greatest possible profit, when production possibilities and prices are given.

8.1. Prices, Information, and Rationing

As we have pointed out in previous chapters, prices carry information. When we know the prices of the various goods and services, it will be easier to compare alternative uses of money. And when it is easy to compare alternatives, it will be simpler to make choices, i.e. to make economic decisions.

Every consumer—in a planned economy as well as in a market economy—experiences the importance of prices in economic choices. Whether to spend 100 roubles on a pair of shoes or on a shirt is an example of a choice we all recognize. Should you choose to spend your money on shoes rather than on a shirt, the cost of these shoes may be expressed in two ways. The usual way is to say that the shoes cost 100 roubles. However, it is also possible to say that the price of the shoes was the shirt you no longer could afford to buy. Expressed in this way, the price or cost of acquiring a good is the best alternative you must do without. This is known as *opportunity cost*.

Before consumers make their purchases, continual trade-offs will be made between all the existing possibilities. Based on the money available and the prices that are in effect, the choices of what you will end up buying will be made. The difference between a consumer in a planned economy and one in a market economy is that the 'planned consumer' must often stand in a queue to obtain goods; in addition to the price in roubles, he or she will have to pay a price in the form of elapsed queuing-time.

In a market economy, prices in money are the only significant thing determining your choice of purchase, given your income. We then say that the price alone functions as a *rationing mechanism* for the scarce goods supplied by the economy. In a planned economy, where prices are set lower than is necessary to clear the market, there will be a need for a further mechanism to distribute the goods. The most common one is undoubtedly the time spent queuing. However, other mechanisms will also be found—for example, special shops where only higher party officials, etc., have access, but where prices are the same as in ordinary shops, and where queues seldom arise.

8.2. Relative Prices and the Absolute Price Level

When the market is to replace planning, prices will be considerably altered. In this connection it is important to distinguish between changes in relative prices and changes in the absolute price level, as we did in Chapter 5. By *relative prices* we mean the number of units of a product one must forsake in order to buy one unit of another product. When the price of a pair of shoes and that of a shirt is 100 roubles, the relative price will be one shirt per pair of shoes. If shoe production was strongly subsidized under the planning system, it is probable that shoes will be relatively more expensive when planning is replaced by the market and subsidies are removed. Free competition may lead to the price of shoes, measured in shirts, rising from one shirt to two shirts. That would mean a doubling of the price of shoes (measured in shirts), or a halving of the price of shirts (measured in shoes).

The transition from central planning to a market economy will quite surely lead to significant changes in relative prices. The acceptance of new relative prices is a challenge that cannot be avoided when the market economy is introduced.

By the *absolute (or general) price level* is meant the average price of all goods and services in the economy. In order to calculate this average, it is necessary to construct a *price index*. This is done by weighing together the prices of the various goods and services. The weights reflect the share of each good in total consumption; cf. Supplement 5.1.

An increase in the absolute price level may be more difficult for most people to accept than changes in relative prices. When the absolute price level rises, there will of course be a reduction of the purchasing power of money as well as of current income. With the great build-up of banknotes and bank deposits that has taken place in many planned economies, a one-time leap in the general price level is unavoidable. We will return to this question later.

To round off the shirt/shoes example at this juncture, we can imagine that with the market economy in place the price of a shirt will be doubled to 200 roubles, while the price of shoes will be quadrupled to 400 roubles. The relative price of one pair of shoes will increase from one shirt to two shirts. And the absolute price level, measured by the number of roubles that must go towards the purchase of one pair of shoes and a shirt, will be tripled, from 200 to 600 roubles.

Thus, there is not a great difference between a planned economy and a market economy as far as consumer behaviour is concerned. Under both systems, the consumer tries to spend money in such a way that the greatest possible utility is obtained, given the existing prices. While prices alone are the rationing mechanism in a market economy, supplementary mechanisms such as queuing are utilized to a varying degree in a planned economy. When prices are the sole rationing mechanism in an economy, we say that the domestic monetary unit (the currency) is *internally convertible*.

8.3. Challenges to Producers

As far as the behaviour of producers is concerned, however, there is an ocean of differences between the planned and market economies. In a planned economy producers are subject to the overall plan devised by the central authority. Their primary task is to fulfil the target figures for physical production. Prices will play a totally subordinate role. So will the enterprises' ability to pay for inputs. When everything is subject to physical planning, there will be no room for any independent role for prices. Firms that get into difficulties because bills cannot be settled will be 'saved' by the state through special transfers; for example, they will be allowed to raise their prices or will be exempted from paying tax. In other words, they work under a 'soft' budget constraint.

In a market economy, on the other hand, prices play a corresponding role for producers as they do for consumers. For producers, too, prices are carriers of information. When there is proper competition, the individual producer will have to take all prices as given.

In our simple shoe example, all the producers face given prices for leather, thread, buckles, rubber, and other production components. Furthermore, the price of labour, i.e. wages, as well as the price of capital, i.e. the interest on bank loans, will also be determined in the respective market. This means that the producer will have a problem analogous to that of the consumer; while the consumer seeks to achieve the greatest possible utility from his money for given prices, the producer or owner of the enterprise seeks to achieve the greatest possible return on his invested capital. In other words, he desires the greatest possible profit, at given prices.

In the same way as the consumer weighs up the various alternative ways of utilizing his money when he is to make a decision about what to purchase, the capital-owner will do the same thing when he decides what he is going to produce and how.

Let us take as a point of departure a person who has 1 million roubles at his disposal, and who is considering establishing a new factory. The money will be sufficient for either a shoe factory or a shirt factory. He has experience from both of these branches, but he is uncertain as to which project he should choose. Thus there will be a need for precise assessments and calculations. First, the technical ones. What kinds of buildings, machinery, and employees would be needed for each of the two projects? How large a production of shoes and shirts respectively could be expected?

The aspiring factory owner must acquire appropriate information, and perform his calculations and assessments, in order to come up with a comparison of the two options. Let us assume that the result of the calculations is an expected annual profit on shoe production of 300,000 roubles, compared with only 100,000 roubles on shirt production. With an investment of 1 million roubles, these estimates would mean a profit on the invested capital of 30 and 10 per cent, respectively. If both projects were considered to involve the same risk, the capital owner would obviously choose to initiate the production of shoes.

We could say that the costs relating to the shoe project are 1 million roubles. Equivalently, we could say that the opportunity cost is reflected in the alternative he must reject, i.e. the shirt project. In the same way as consumers must consider alternative ways of utilizing their money, with a view to maximizing utility, the capitalist must do the same, with a view to maximizing profit.

Let us now assume that the shoe-manufacturing enterprise is established and the enterprise-owner's calculations are initially correct. Then he will be able to enjoy a very good profit on his money. At the same time, others may easily come to envy him and find it unjust that one person should earn so much on other people's work. And here we have arrived at one of the market economy's crucial dilemmas: namely, the imbalance in income distribution which may easily follow.

Income inequality can, however, be modified by means of a progressive tax system, whereby the rich contribute proportionately more than the rest to the financing of the public sector; also, the less well-off benefit relatively more than others from public expenditures and from various types of transfers such as welfare payments in cases of illness or unemployment, scholarships, and cheap student loans.

We can go one step further in our example. The fact that profit-making prospects in connection with shoe production are so good would indicate that society has too small a supply of this good. When production costs are 300,000 roubles less than sales revenues, this means that society's willingness to pay is significantly higher than resource consumption in this industry. The high profit can quite simply be interpreted as an expression of the fact that total shoe production in society is too low and accordingly should be increased. Against this background, we could say that the capital-owner is only doing his duty when he initiates such production.

8.4. New Profit-Making Prospects

These considerations bring us to another challenge in connection with the transition from plan to market: namely, that producers must demonstrate a capacity to assess and react to profit opportunities. In this way they will participate in producing those goods and services that society most lacks. An important prerequisite for a successful market economy is therefore the existence of *supply response* among producers, meaning that they must be able to respond to the new profit-making prospects that arise.

If, in the example we have looked at, there is a strong supply response, the capitalist who is first to initiate production of shoes will earn good money. (We let him have 30 per cent profit initially.) However, when enterprising persons see the golden opportunity for profit in the production of shoes, more of them will join in. Production of shoes will increase, competition will become more intense, prices will be forced down, and profit margins will decrease. For consumers this will be good news; the supply of shoes will increase and prices will fall.

In the long term, profits in shoe production will tend to approach the same profit levels as those in the rest of the economy, for example in the production of other goods (see Supplement 8.1). If too many initiate the production of shoes at the same time, prices may be forced down so far that the least effective firms will run at a loss. After a while the bankruptcy mechanism will come into force. Those enterprises that survive will then have better prospects inasmuch as profits will begin to rise again.

SUPPLEMENT 8.1. OVER A PERIOD OF TIME PROFITS IN ALL INDUSTRIES WILL TEND TO APPROACH A NORMAL LEVEL

In order to illustrate the development in the shoe industry that is described in the text, we can utilize a diagram. In Fig. 8.1 we have quantity produced on the horizontal axis and price on the vertical axis. As in Chapter 3, the demand curve slopes downward. This means that an increase in price will reduce demand. On the supply side it is the opposite: when the price increases, producers will wish to supply a larger quantity.

Suppose the economy is in equilibrium at point A. With this relatively high price for shoes, several new shoe factories will be established. With more producers and increased production, the supply of shoes will increase. This will lead the supply curve to shift further out. But with increased production, the price will have to be reduced. The new point of equilibrium for the market will be at B. Compared with point A, production will have increased and the price will have fallen. Not surprisingly, profits will also have decreased. The unusually high profits that the shoe manufacturers were able to enjoy were thus merely a temporary phenomenon.

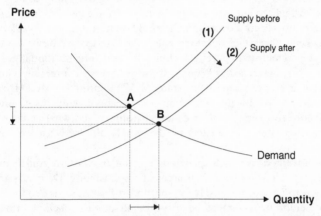

FiG. 8.1. More producers emerge and the supply curve shifts further out

By means of this simple discussion of shoes and shirts, we have sought to illustrate Adam Smith's 'invisible hand'; through his continuing search for the best profit-making prospects, the capital-owner does his job. For in areas where profitability is particularly high, society's need for more goods is especially great. Over a period of time capitalists will spoil things for one another; through competition, profits in the various branches will be forced down to a normal level, which, corrected for risk, will tend to be the same throughout the economy.

The producer and the consumer both measure alternatives against one another and choose what seems to be best. If the prices in the economy correctly reflect the scarcity of the various goods and services in society, those decisions that are made on this basis will be 'good' (more on this at the end of the chapter). By this is meant that production becomes more effective at the same time as it satisfies the needs that people in society, through their demand, show that they have.

So far in this chapter we have focused mainly on people's behaviour as consumers and as capital-owners in a market economy. We shall now examine the challenges firms face when the market economy replaces the planned economy. Thereafter we will discuss the challenges that have to do with mentality of people, including their attitudes towards free enterprise. To conclude the chapter, we shall discuss some challenges to the authorities.

8.5. Reorganization in Enterprises

The market economy is not a bed of roses, for either the capitalist or the worker. The struggle to survive, and preferably to make good money, will place demands on all employees. In the individual enterprise, it will be necessary to keep abreast of developments on a day-by-day basis: could production be structured purely technically and organizationally in another way, with lower costs? What enterprises are we competing with, and how have they prepared their strategy? What products should we concentrate on in the future, and which ones are making losses and should be axed? How satisfied with us are customers, and what could we do to make them more interested in the goods and services we can supply?

In a well-run enterprise such questions are under constant consideration. Managers at all levels have their areas of responsibility. Demands are made on them to submit proposals for solutions to the various problems. The task of senior management is to weave all of this together into a united strategy for the enterprise. This will provide a basis for concrete plans that managers at lower levels must implement.

For enterprises that are used to being informed by a central authority of what is to be produced, the transition to a market economy will involve great challenges. No longer will it be sufficient to sit passively waiting for directives from above. Instead, it will be necessary for the manager of a firm to take the initiative in analysing and evaluating both production and market prospects, and to prepare plans of action in collaboration with other managers in the enterprise.

Implementation of such plans will obviously affect individual employees. When the composition of production is altered or when new technology is brought into use, many will get new work assignments. This will require flexibility and continual further training of the individual worker.

On the other hand, a successful transition to a market economy will result in substantial growth in production and in the enterprise's wage-payment capacity. More goods will be available for distribution, and most people will experience an improvement in their material living standards. The individual will also attain more freedom concerning choice of residence, education, and occupation. As an employee he will have to measure these advantages against greater demands on work contribution and less job security.

In a market economy, too, there is a large public sector with many jobs. In this public sector people very seldom lose their jobs. This is because the bureaucrats and others who work in this sector are not subject to the harsh law of the market. Put another way, ministries, directorates, and other public agencies cannot go bankrupt. The disadvantage for public-sector employees is lower pay than they would get for corresponding qualifications in private enterprises. So the relative lack of job security in private enterprises or enterprises exposed to competition is compensated for in the form of higher pay.

In a market economy much time and effort is spent on investigations of how production, sales, etc., can be improved. However, unlike in a planned economy, the individual enterprise is independent of the authorities. Planning within the enterprise is aimed at doing as well as possible in the face of competition. And continually hanging over the enterprise will be the Sword of Damocles: should it fail, the capital-owners may see their money vanish and bankruptcy and closure may become a reality.

In industries where things change rapidly, be it new technology, new competitors, or continual changes in consumer preferences (fashions), the demands on adjustment in the individual enterprise will be particularly great. And in contrast to the conditions under a planned economy, enterprises that run into difficulties cannot count on support from the authorities.

Let us now go on to consider the challenges that are more generally related to people's mentality and attitudes.

8.6. Changes in Mentality and Attitudes

As we have previously observed, a market economy will provide an opportunity for people who are both proficient and lucky to become rapidly wealthy. Those differences in income which will then emerge, and which in a way are the driving force of capitalism, can be reduced somewhat through various tax and welfare schemes. However, the fact that significant and open differences in income and living standards will arise cannot be avoided when a market economy is introduced. For most people who do not belong to the 'nouveaux riches', accepting this consequence of the market economy may prove to be a challenge.

A further comment might be appropriate here. Skilful capitalists, who establish profitable enterprises, make a contribution which benefits society as a whole. The wealth that they accumulate for themselves will thus proceed from something that benefits more people than themselves. In a planned economy, on the other hand, there are far too many examples of party bosses who have obtained privileges and economic advantages at the expense of others, while their contributions have *not* enriched society—on the contrary. Thus, in a market economy there will be a sharper distinction between the economic sphere and the political one than in a planned economy. Indeed, this is one of the prime advantages of capitalism. Against this background, one might perhaps expect that differences in incomes would be easier to accept under a market economy than under a planned economy.

Another challenge is the acceptance of the market's impersonal nature. Prices 'come into being' in a market characterized by competition. The individual producer has little latitude in his price-setting. If he sets his price higher than his competitors, his sales will fall sharply. And if he reduces his price below what the market will bear, he will lose money: after all, he could have sold the same quantity at a higher price.

In connection with bankruptcies, too, the impersonal nature of the market enters the picture. When an enterprise no longer has sufficient liquid resources to pay its bills, it will 'go' bankrupt. All that is needed is that a creditor send his claim to the probate court. No active decision by a public authority is required.

In the public sector, on the other hand, even in a market economy, 'someone' must take explicit decisions if the closure of undertakings is to take place. We could mention hospitals as an example. In Scandinavia virtually all hospitals are publicly owned and operated. In the last couple of decades it has become widely recognized that many hospitals operate inefficiently. For example, the period of hospitalization and utilization of resources in connection with the same type of operation may be twice as much at a poorly run hospital compared with a well run one. If the law of the market had applied in this case, the less efficient hospitals would have been bought up and reorganized by a competitor, or forced to close because of their lack of competitiveness.

However, the law of the market does not apply to Scandinavian hospitals. And there may be good reasons for that. The relevance of this example for us is that closure of a hospital or section of a hospital will require an unpopular political decision. And the politicians who determine these things find it difficult to make such decisions.

In a planned economy it is not only hospitals and the health service that are a public-sector responsibility: the whole of trade and industry is owned in principle by the state and is directly subject to control by the state. For this reason, the bankruptcy mechanism has not been in operation. When the market is to replace planning in large areas of trade and industry, many will have the unpleasant experience of the impersonal mechanism governing closures and reorganization that bankruptcy implies. On the other hand, when everything is governed by the impersonal market rather than by individual heavyweight politicians, there will be less room for nepotism and the random exercise of personal power.

A third challenge concerning mentality of a similar nature has to do with concepts such as independence and freedom. In a market economy the individual has more freedom, both political and economic, than in a centrally directed society. The exercise of this freedom involves independent choices whose consequences one must bear oneself.

For better or for worse, the individual in a market economy is 'a self-made man' to a far greater extent than in a planned economy. If, after finishing your education, you have difficulty in finding a job, no state enterprise will stand ready to employ you. And if you work well in an enterprise that does badly, you may find yourself out of work one day; no paternalistic state or public authority will exist to save any enterprise threatened by bankruptcy. On the other hand, you will receive unemployment benefit and assistance in finding a new job. The safety net provided by the government is thus designed to ensure survival of the individual human being—but not survival of the individual enterprise.

The knowledge that this is so, and the experience that is gradually gained, will probably give rise to greater adaptability to the market's impersonal and often unpredictable manner of operation. Through such a process, the way will be paved for the market economy to function as intended.

SUPPLEMENT 8.2. AMERICAN AND SOVIET ATTITUDES TO THE MARKET ARE NOT SO DIFFERENT

In May 1990 a comprehensive interview survey was carried out in Moscow and in New York concerning people's attitudes to the market economy. The questions were the same: were the answers, too? Surprisingly enough they were, to a large extent.

As far as attitudes to income differences and belief in the importance of reward for effort were concerned, it was hard to trace any differences. It is true that the Soviets' attitude to businessmen was somewhat cooler than the Americans'. But there seemed to be greater understanding of the manner of operation of a market economy in Moscow than in New York. Researchers view this in connection with the ongoing debate on transition from a planned to a market economy in the Soviet Union.

From page 40 in the concluding final section, we have taken the following excerpt:

Because the differences between the USSR and the USA we found were often small or nonexistent, we feel that perhaps too much prominence has been given in discussions of the transition to a market system in the Soviet Union today to the difference between Soviets and people in market economies. The pressing and immediate problems faced in the Soviet Union today may be instead political and institutional in nature. When a country inherits an institutional and political framework that has been anti market, it serves certain entrenched interests in that country to resist change. Thus individuals who benefit from the present system may make public appeals to fairness, abhorrence of income inequality and other attitudes to try to stop change. Alternatively, well-meaning Soviet government planners may feel constrained by their incorrect belief that the Soviet public is much more concerned with fairness or income equality than are the publics in capitalist countries.

Source: Shiller *et al.* (1990)

8.7. Challenges to the Authorities

Finally, we shall say a little more about some of the challenges the authorities will face when the market system comes to replace the planning system. In the last chapter of the book we will be coming back to these difficult and important issues.

As we have seen, a well functioning market will require an appropriate code of legislation which guarantees private ownership. In addition, laws and regulations will be needed to combat tendencies to form monopolies. When supply and demand, under free competition, determine all prices, will the authorities then be able to relax, and let everything function by itself? Of course not.

A classic example of how calculations of economic costs in business enterprises underestimate the economic costs to society is pollution. When an enterprise spews toxic gases into the air or harmful chemicals into the water, it is obvious that society as a whole will suffer. This means that society will have a cost imposed upon it. However, in the absence of taxes or other forms of regulation, the polluting enterprise will not take these costs into account. Free competition, with no attention being paid to the pollution caused by various forms of consumer and business behaviour, will result in erroneous production and consumption decisions. This is due to the fact that the information carried by prices is misleading. In order for this information to be correct, *all* costs, from society's point of view, must be present in calculations, and thus reflected in prices.

When economic behaviour has consequences beyond those that are reflected in prices, we say that *external effects* are at work. (This was mentioned briefly in Supplement 2.1.) In cases of pollution these are negative external effects. However, there are also positive ones. The classic example is that of the man who has an apple orchard which benefits his neighbour who keeps bees and produces honey. Flowering fruit trees naturally provide the best working conditions for bees in their hunt for pollen and nectar, which will be converted to honey.

The problem with external effects in a market economy is that the individual will not take into account the harm or benefit his undertaking causes to others; i.e. the costs and benefits to others will not be reflected in his prices. Thus, the prices will not be carriers of correct and complete information.

One might believe that in a planned economy conditions are more favourable for avoiding major problems with external effects. In development of the overall plan, which encompassed all physical production, it ought in principle to be possible to take into account the advantages the fruit-farmer creates for production of honey; since there were positive external effects, production of fruit ought to be greater than the level indicated by private economic calculations. In the fields of art and science there are examples of planned economies actually having taken such positive external effects into account. However, when it comes to the negative external effects of pollution, the record of planned economies is dreadful. Inexpensive energy has been overused, and so have old-fashioned, unclean methods of production (see Supplement 8.3).

SUPPLEMENT 8.3. PLANNED ECONOMIES AND NEGATIVE EXTERNAL EFFECTS

The Central and Eastern European planned economies have on the whole been much less successful in dealing with negative external effects, such as pollution, than the Western European market economies. Despite a far lower level of total production, pollution in many places in Central and Eastern Europe has reached catastrophic dimensions.

In this connection, it is enough to mention great inland seas in the former Soviet Union, such as the Aral Sea, and a city such as Cracow in Poland. When water from the Aral Sea was used for irrigation of cotton fields, the water level fell and the salt deposits were revealed; the salt blew over the cotton fields, which became desolate in time. In Cracow the extensive use of brown coal, with the corresponding emission of sulphurous gases, has resulted in noxious air and in buildings gradually disintegrating in consequence.

The enormous pollution problems plaguing Central and Eastern Europe may be regarded as evidence that the system of economic planning in practice is something quite different—and less pleasant—than the way planned economies were meant to function in theory.

In a market economy one could imagine solving the pollution problem by means of an extension of ownership rights. Let someone—for example, the authorities—own the clean air and the pure water. Pollution-causing production would in such a case mean that the producer would have to buy air and water before he could pollute these previously free goods. But when the pollution costs something, the profit-maximizing enterprise-owner would have an incentive to reduce emissions himself. It would become *profitable* to reduce emissions—for example by installing cleaning filters. In Chapter 10 we shall take a closer look at how the market mechanism can be used effectively to solve the problem of pollution. However, first we shall study some problems that are more specific to a market economy.

9

Dilemmas and Problems in a Market Economy

> Industrial crisis, unemployment, waste, widespread poverty, these are the incurable diseases of capitalism.
>
> Joseph Stalin

ALTHOUGH the 'competition' between the planned and market economies as economic systems now seems to have been decided in favour of the market, a decentralized market economy is not without dilemmas and problems. In this chapter we shall examine some of them, against the background provided in Chapter 5. However, let us first look at some history.

9.1. The Great Depression and Keynes

In the 1930s, countries such as the United States and the United Kingdom experienced unemployment rates of over 20 per cent for several years; that is to say, two out of ten people who wanted a job could not get one. In parts of Europe, as much as one-third of the labour force was out of work for years. Immense unemployment entails a very substantial loss of production, and also causes great human suffering, both material and mental.

While the 1930s are usually associated with the 'Great Depression' in the West, this was a period of formidable industrial growth in the Soviet Union.

SUPPLEMENT 9.1. THE 1930S: PLAN SUPERIOR TO MARKET?

From 1928 to 1931 unemployment in the United States rose from under 4% to over 25% of the labour force (see Fig. 9.1). The market economy was in serious trouble. And ten years would pass before unemployment reached a normal level again: 6% in 1941. (In the war years 1942–5 the American economy was characterized by 'overfull' employment; i.e., many more than normal were in paid employment. For this reason there are no unemployment figures for these four years in Fig. 9.1.)

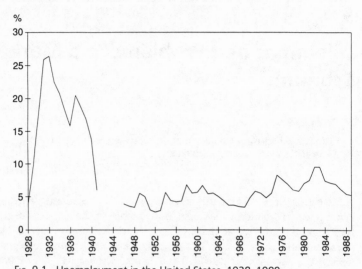

FIG. 9.1. Unemployment in the United States, 1928–1989

Sources: ILO, *Yearbook of Labour Statistics,* 1928–1964; OECD, *Labour Statistics,* 1964–1989

From the end of the war and until 1991, the unemployment rate in the United States varied between 2.7% (1955) and 9.5% (1982 and 1983). With an unemployment rate of more than 6%, this country is wasting its most important economic resource: people who want to work. In addition, high unemployment creates serious social problems.

Fig. 9.2 shows developments in Soviet and American industry for the years 1928–36. With an index of 79.5 in Soviet in 1928 (and 100 in the base year 1929), the index stood at over 380 in 1936! While the United States was struggling with an average unemployment rate of just over 20% in the years 1928–36, and a significant fall in industrial production, in the same period the Soviet Union was enjoying an annual growth in industrial production approaching 22%. Towards the end of the 1930s, the system of economic planning appeared to be a serious challenge to the market economy.

However, we must not forget the reverse side of the coin. The rapid growth in Soviet industrial production illustrated in Fig. 9.2 took place at the same time as Stalin's gruesome collectivization of agriculture and the establishment of a large number of work camps. Millions of people were sent by force to these camps, and many died because of the treatment they were exposed to there.

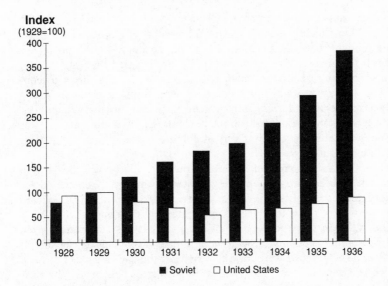

FIG. 9.2. The index of development of industrial production in the
Soviet Union and the United States, 1928–1936

Note: The index = 100 in each country, 1929.

Source: Statistical Yearbook of the League of Nations 1937/38, Geneva, 1938

In academic debate at the time, the Polish economist Oskar Lange
maintained that a planned economy would be better able to utilize
resources than a market economy. By means of continually changing
prices stipulated by the central authority, planning could *imitate* the market. In addition, planning would ensure full utilization of all resources in
the economy.

In contrast to Lange, the Austrians Friedrich von Hayek and Ludwig
von Mises asserted that such an imitation would not work. The absence
of the profit motive, the enormous demands regarding information concerning production and demand conditions, and the lack of political freedom that the individual would have to accept would eventually result in a
command economy with no capacity for renewal or growth. Today Hayek
and Mises are being read with renewed interest.

However, the economist who was of greatest importance to the development of the capitalist system from the mid-1930s up to the end of the
1970s is undoubtedly the Englishman John Maynard Keynes. Keynes was

concerned about the acute problems of his time, first and foremost the high and long-lasting unemployment. While most of Keynes's contemporaries thought that over a period of time the economy would move back by itself to a state of full employment, Keynes claimed that this was not necessarily the case. If the willingness of households and firms to consume and invest was small, the aggregate demand in the economy would be less than required for full employment. In such a situation the authorities ought to intervene and stimulate demand. In many cases it was Keynes's opinion that the best way of doing this was to increase government expenditure.

Keynes's thinking gradually gained ground, first among colleagues in academia, subsequently among the politicians who, partly on the advice of academic economists, converted Keynes's ideas into practical policies. The 1960s were the golden age of 'fine-tuning' of the national economies of the West. In this decade most Western economies experienced even and substantial economic growth while inflation was held reasonably in check. The economists were the high priests of the day. They had 'broken the code'; from now on, lean years and unemployment would be relegated to the history books.

9.2. Inflation and Unemployment

However, Adam did not remain in Paradise. In the United States demand was maintained over a long period of time at a higher level than was necessary to ensure full employment. This was connected with the fact that President Lyndon B. Johnson waited too long before raising taxes to finance his two major programmes: the war in Vietnam, and development of the United States into a modern welfare state. Rather than asking Congress to increase taxes in order to finance the increase in public expenditure, he asked the Federal Reserve (i.e. the central bank) to print more money. Initially this resulted in higher wages and increased purchasing power for most people. Subsquently it led to increased inflation: the rate of inflation, which was well below 2 per cent in the United States in 1960, had risen to almost 6 per cent by 1970.

At the beginning of the 1970s the Organization of Petroleum Exporting Countries (OPEC) flexed its muscles, and quadrupled the world market's price for crude oil. The increased oil prices involved a formidable transfer of purchasing power from the oil-importing countries to the oil-exporting ones. The Western economies entered a period characterized by rising unemployment and rising inflation.

SUPPLEMENT 9.2. FROM THE PHILLIPS CURVE TO STAGFLATION

Up to around 1970, experience had shown that inflation and unemployment seemed to be inversely related to each other in the United States, the United Kingdom, and many other market economies. This meant that lower unemployment could be 'bought' by higher inflation; in other words, by stimulating aggregate demand the government could reduce unemployment at the cost of higher inflation. This inverse relationship is expressed in the Phillips curve, and is shown in Fig. 9.3(a). (The curve is named after Arthur W. Phillips, an economist from New Zealand who discovered the inverse relationship; see Phillips 1958.)

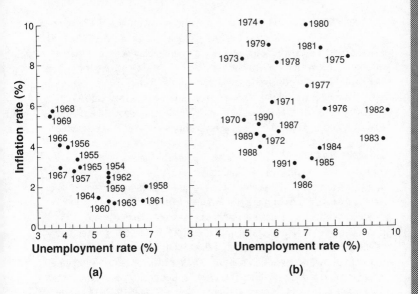

FIG. 9.3. Phillips curves for the United States, 1954–1969 and 1970–1991

Note: Unemployment is measured by the commonly used definition. Inflation is the percentage change in the private consumption deflator from the previous year.

Source: OECD World Economic Outlook, 50 (1991): Tables R11 and R19.

This phenomenon can be explained as follows. When aggregate demand is great, virtually all resources in the economy will be fully employed. As far as labour is concerned, this means that by and large all who want to work can get jobs. A high level of aggregate demand for goods and services will thus result in low unemployment, other things being equal. On the other hand, demand pressure will make it easier for firms to raise their prices. If their managers can sell their whole production this year at a price that, for example, is 10 per cent higher than last year, they will—as good businessmen—do exactly that. Thus, a high level of aggregate demand also leads to increasing prices, that is, inflation. For example, 1969 was a year of low unemployment and high inflation by historical standards in the United States, as can be seen by the Phillips curve in part (a) of the figure.

A decrease in aggregate demand for goods and services will make it difficult to raise prices, and will also dampen the demand for labour. Thus, low inflation will tend to be associated with high unemployment in this case, other things being equal. This is illustrated by the point for the year 1961 in part (a).

In the 1970s and 1980s the simple relationship between unemployment and inflation, as expressed in the Phillips curve, disappeared. We got the worst possible scenario: both inflation and unemployment rose sharply (see part (b)). A single downward-sloping curve which would fit the data, as in part (a), is impossible to draw through the data points in part (b). The 'stagflation' of the mid-1970s and early 1980s seemed shrouded in mystery for a while. However, it is not difficult to understand against the background of Chapter 5. There we saw how an increase in production costs—triggered, for example, by a quadrupling of oil prices in world markets in 1973–4 and a doubling in 1979–81—leads to an increase in inflation and unemployment simultaneously.

At the same time, it became widely recognized that persistent inflation creates expectations of more of the same. In order to come down to moderate inflation, it was necessary to break these expectations. This happened in the United States in the first half of the 1980s, when the Federal Reserve (the central bank of the United States) reduced the rate of growth of the money supply substantially in order to restrain aggregate demand and reduce inflation. This worked, but at a price; the unemployment rate went up to almost 10 per cent in 1982 and 1983. Thereafter, unemployment was gradually reduced to below 6 per cent in 1988–90. A similar story can be told about the United Kingdom and several other countries in the 1980s.

In an attempt to resolve this dilemma, President Richard M. Nixon introduced strict regulation of wage and price developments between autumn 1971 and spring 1974. For a short period his administration quite simply prohibited enterprises from increasing prices and wages. Then the government went on to issue binding guidelines concerning wage increases, before these measures were gradually phased out.

Although regulation of wages and prices was a popular measure—most people liked the fact that the President acted firmly in a difficult situation—this policy was without any long-term effect. Before the regulations were finally phased out, inflation had begun to rise. And by the mid-1970s annual inflation in the United States had risen to over 10 per cent (1974) while unemployment was running at more than 8 per cent (1975) (see Supplement 9.2).

Varying and at times high unemployment is something a market economy can never get rid of entirely. In the short term there may be a conflict between the desire for low inflation and the desire for low unemployment. In the long term increasing prices will become an integral part of people's expectations and will be reflected in wage demands. The will to pay higher wages will also increase when enterprise managers count on the fact that they will always be able to raise the prices of what they produce.

Over time, unemployment will tend to approach what is called 'the natural rate of unemployment' for the economy. By this is meant the rate of unemployment at which there is no tendency for inflation to increase. This explains why some unemployment—presently, about 6 per cent of the labour force in the United States—is considered 'natural' and consistent with 'full' employment. If the authorities take it upon themselves to reduce the natural rate of unemployment without increasing inflation, focus must be shifted from the demand side of the economy to the supply side, as we discussed in Chapter 5. For example, schemes for retraining the unemployed may have a positive effect in this situation. Further, it may be necessary for wage levels to be reduced for those occupational groups where unemployment is particularly high. Chapter 14 deals with the labour market, and there we will look more closely at some of these questions.

9.3. Income Distribution and Luck

High unemployment brings us to another problem with a market economy: namely, the imbalance in income distribution. We have previously pointed out that 'hard' budget constraints and bankruptcy enable a market economy to reorganize efficiently. However, we have also stated that this flexibility in the system may create winners and losers in a random and unjust way. High unemployment is therefore not merely a problem in the sense that the country's total production is less than it might have been: it is also a problem from the point of view of how what is produced is distributed.

In addition to proficiency, good luck and bad luck are a part of the picture. For example, take Central or Eastern European students who have studied English diligently in recent years. Then comes *glasnost* and openness towards the West. The value of that knowledge will be far higher than could have been imagined when they began their studies. The prospects of interesting and well-paid jobs in enterprises that are orienting themselves in the direction of the market economies in Western Europe and the United States will be legion. On the other hand, those who finished their economics studies in the mid-1980s, and thus only learned how a planned economy works without gaining insight into the workings of the market economy, will have knowledge of little relevance to offer in the labour market of the 1990s.

9.4. Equality and Efficiency

In spite of the fact that at the beginning of the 1990s unemployment in the United States was at the same level as ten years before, just under 6 per cent, differences in income have grown significantly. A recent study concludes by saying that average real income before taxes among the 10 per cent wealthiest families increased by 21 per cent from 1979 to 1987. For the 10 per cent most disadvantaged families in the same period, there was a *reduction* of 12 per cent. The growing gap between rich and poor is due to several factors: growing differences in income between skilled and unskilled workers, increased labour force participation of women in wealthy families, and something that must be called a game of hazard in financial markets in the 1980s.

In order to solve the problem of poverty in the United States, it is necessary to introduce far greater transfers of purchasing power to the poor without reducing their incentives to work. In addition, there is a need for an increased public-sector contribution to the improvement of housing and education for the disadvantaged. With President Ronald Reagan in the driving-seat, however, there was no political will to implement such a policy in the 1980s. The philosophy of his administration aimed rather in the opposite direction: at the reduction of taxes and public expenditure.

In Western Europe, particularly in the Scandinavian countries (Norway, Sweden, and Denmark), the equalization of incomes and wealth has gone significantly further than in the United States, even though in these countries, too, there are great differences between the wealthy and the less well-off. The 'Scandinavian model' for a more egalitarian society, however, is not without problems. This is because the incentive to work hard and make risky investments lessens when the authorities tax 'winners' and subsidize 'losers'. Such a policy weakens the link between performance and reward.

Here we can perceive a dilemma inherent in the market system: if we seek to ensure roughly the same-sized piece of cake for everyone, we may not be able to have the biggest possible cake for sharing. Establishing the trade-off between equality and efficiency is a problem which the market economy will have difficulty in resolving.

Under the system of economic planning, equality has in principle been ranked in front of efficiency. In certain quarters—not least among the old powers that be—the desire for relatively small differences in income has led to an interest in the 'Scandinavian model' in Central and Eastern Europe. But in an economy which, in its efforts to gain more from its resources, chooses to gainsay planning in favour of the market, it is not possible to be too ambitious in this respect. If the market is to work according to intention, proficiency and luck must be rewarded. And the pressure on the individual to do his best must not be removed.

An example will illustrate the point. Imagine that the authorities in Lithuania, wanting all the inhabitants of the country to have reasonably good housing standards, decided that a family of four persons with a monthly income of 1,600 roubles or less should receive 100 roubles per month in housing benefit. The intention would be the best. But how would such a scheme work? Say that a family in Lithuania received 1,600 roubles in monthly income, so that they were entitled to the 100 roubles as housing benefit. If the prospect arose of their working some more, so that their income rose to 1,700 roubles, would the family do so? Hardly. The reason is that the 100 roubles the family received in additional income would correspond exactly to the 100 roubles they would lose in housing benefit. The economic reward for the increased work effort would be equal to zero. And if the family were rational in their economic assessments, they would decline the offer of additional work.

The point is that well-meaning welfare schemes often serve to reduce the individual's productive contribution to the economy. The incentive effect of different subsidization schemes must therefore be included in the evaluation of such schemes. For an economy making the transition from plan to market, there will be a need for a coarse-meshed safety net, with the intention of guaranteeing a minimum living standard for everyone in the country. Detailed assistance schemes based on housing standards, transport costs, and the like will not be affordable before total production in society has risen to a significantly higher level.

In the Scandinavian countries there has been increasing recognition of the negative incentive effects of excessive government intervention in economic affairs in general. Since the early 1980s great efforts have been made to change the system of taxation and welfare transfers so that the incentive for the individual to make a productive contribution will be greater. This has resulted among other things in extensive reorganization of the tax systems in Sweden and Norway. Further, in Sweden wage payments in connection with short-term sick leave have been reduced—with the result that the population has suddenly become much healthier.

9.5. Stupid Capitalists

In order to understand the way a market economy works, we have previously described the market's central participant, 'Economic Man'. He is expected to make rational calculations and, based on existing prices, to choose those options that give him the best result. His expectations concerning the future must also be contained in these calculations. As a consumer he must assess both current income and what he expects to earn in the future; as an investor or capital-owner he must form a picture of the prices he can count on obtaining for his future production.

In the last chapter we let the investor have a 30 per cent profit on shoe production in the first year. However, the story continues; with such a good profit he must expect that more people will want to get in on the act. This means stiffer competition and reduced profits. (See Supplement 8.1, where the supply curve shifts outwards.) In his assessment of the shoe-manufacturing undertaking's profitability over a period of time, therefore, it would be wise for the capitalist to assume that a profit of 30 per cent is hardly going to continue indefinitely.

Of course, it is never easy to know what the future will bring. Nevertheless, it is difficult to avoid the conclusion that capital-owners in a market economy now and then behave more like a flock of sheep than like rationally calculating 'Economic Men'. Especially in the financial markets, we experience from time to time speculative 'bubbles' that swell up, without any long-term economic foundation, before finally bursting.

A classic example is the American stock market crash in 1929, when on one day in October prices on the New York Stock Exchange fell by almost 10 per cent. And then, one day in October 58 years later the stock market index in New York fell by more than twice as much—and in many countries by even more. In both cases share prices had risen to totally unrealistic levels, driven by something that can hardly be called anything but the 'flock of sheep mentality' among capital-owners.

However, in the real estate market, too, there occur examples of exaggerated and groundless optimism from time to time leading to unreasonably high prices and far too bold investments. For the economy as a whole, the problem of overinvestment within an industry constitutes inappropriate utilization of resources. As an illustration, we shall look briefly at developments in Norway in the latter half of the 1980s.

Normally 95 per cent of office and shop space is in use in Norway; this implies that the rate of vacancy is 5 per cent. This vacancy rate is due to the fact that there are always some enterprises that are cutting back or closing down their operations; their spaces will become vacant until the premises are again put to use by new lessees. In the years leading up to 1988, the rental for office and shop premises had risen more rapidly on average than the general price level. This meant that those who owned such premises were earning more and more. This encouraged new building. However, there can be too much of a good thing. By studying readily available information it was possible early in 1988 to see that, with all the buildings that were under construction, there would be a 15 per cent surplus supply of premises at the end of the year. This would lead to reduced rentals.

None the less, optimistic investors initiated more and more building projects. And when the authorities introduced restrictions on the construction of new commercial buildings in Oslo, the capital, many investors were greatly annoyed.

The restrictions came too late and had a minimal effect. The result of the over-large investments was that in the course of two years the price of real estate in Oslo fell by almost 40 per cent. Many investors went bankrupt. And the banks, which had lent them funds, had to write off enormous sums. The crisis in the real estate market thus led to a crisis in the financial sector.

When labour, materials, and capital are utilized to build premises no one wants to use, society will suffer an economic loss. For the Norwegian economy as a whole, decisions made by stupid capitalists, with the willing assistance of stupid bankers, led to the wasting of society's resources.

When capitalists' assessments of the economic prospects are deficient and characterized by the 'flock of sheep mentality', the result will be poor. On the other hand, the example of the real estate market in Oslo shows the robustness of the market. Capitalists *do* go bankrupt, banks *do* take losses, and investments in new buildings *do* come to a halt after a while. The resulting lower rentals lead to increased demand for office and shop premises. After a few years things will undoubtedly be back to the normal vacancy rate of 5–6 per cent. Then once again there will be a reason for increasing the tempo of constructing new buildings.

In this concrete case it could be claimed that the authorities should have intervened, both earlier and more firmly. However, if the authorities, every time they suspected something was wrong, were to intervene with detailed regulations for the economic dispositions of private investors, we would undermine the market's function. Such conduct on the part of the state would create inappropriate expectations among the private capital-owners; if things went well they could keep the profits themselves, while if they went wrong the state would come in and take over. In such a case we would be approaching the 'soft' budget constraint of the planned economy. And over a period of time, we know that this system does an even poorer job of utilizing economic resources. The problem of stupid capitalists can hardly be solved by intervention by the authorities. After all, we have no guarantee that politicians and bureaucrats in general are any smarter, even if they were so in connection with investments in commercial premises in Oslo in the latter half of the 1980s. Since capitalists invest their own money in the various projects, it is reasonable to expect that in spite of everything they will make a greater effort to ensure good profitability than the run-of-the-mill bureaucrat would.

9.6. Market Imperfections

In general terms, we could say that a primary duty of the authorities in a market economy is to ensure that prices reflect to the greatest degree possible the relative scarcity of the various goods and services. If there is a producer monopoly this will not be the case; production will be too small and the price will be too high. In the case of pollution with no form of regulation, things will also go wrong; the environment will be damaged more than is acceptable.

When prices do not reflect relative scarcity, the market has failed to do its job. The resulting market imperfections may be grounds for the authorities to intervene. In the case of monopolies, it may be advisable to pave the way for the establishment of new firms, or for competition through imports from abroad. In connection with pollution, the authorities can use taxes to make sure that polluters pay their way or clean up their act (more on this in the next chapter).

However, there exists the danger that well-meaning politicians will be too eager in their efforts to correct faults in the market. In such cases minor imperfections in the market may be replaced by major ones implemented by bureaucrats and politicians. Thus, against an imperfect market one must weigh imperfect regulations.

Those who benefit from the erroneous prices that market imperfections lead to will normally struggle to retain these benefits. Both the monopolist and the polluter will point to the danger of job losses if the authorities liberalize imports or impose taxes on pollution-creating production. In such cases the employees in the enterprises affected will often take a common stand with the owners and attempt to pressurize the authorities into letting everything continue as before.

For the economy viewed as a whole, market imperfections are an expensive luxury. If such problems are to be solved, there will be a need for politicians who have both an insight into economic relations and the courage and strength to allow the interests of the community at large to take priority over pressure groups.

10

Why the Market Nevertheless is Useful

The best economic system is the one that supplies the most of what people most want.

John Kenneth Galbraith

IN Chapter 2 we concluded that economic growth has its roots in division of labour and increased specialization. Growth is encouraged via investments and the accumulation of capital.

Despite the fact that planned economies generally seem to allocate more resources to investments in new production equipment than do market economies, specialization under the planned system has not been carried as far as under the market system. In Chapter 6 we pointed out that this is connected with the absence of free trade between enterprises and the uncertainty surrounding supplies of necessary inputs.

10.1. The Market Economy in a State of Continual Change

This brings us on to a third factor of supreme importance to economic growth: the capacity of an economy to ensure satisfactory utilization of existing resources. And here the market, through decentralized decision-making, based on the going prices, has shown itself to be totally superior to the planned economy.

In a competitive market, where everyone is striving to achieve the best possible result for himself, prices are determined through the interplay between supply and demand; they are a result of the millions of decisions that producers and consumers as a whole must make every day. These prices function as guidelines for new decisions. In this way, a mechanism for co-ordination of economic activity in society emerges. This makes changes possible; indeed, it ensures that the desired changes do take place continually.

The secret behind the market's effective capacity for co-ordination, as we have said before, is the role of prices as carriers of information. When prices are 'right' from society's point of view, when private ownership has been securely established, and when all the participants, in free competition with one another, are striving to achieve the optimum result for themselves, then the market will indeed supply a continually increasing quantity of goods and services.

SUPPLEMENT 10.1. POOR ADVISERS?

In 1985 Abel Aganbegyan became the chairman of the Commission for the Study of Productive Forces and Resources and leader of the Economics Section of the Academy of Sciences of the USSR. In 1988 he published the book, *The Challenge: Economics of Perestroika*. On the front cover of the book he is designated 'Chief Economic Adviser to Mikhail Gorbachev'.

This book makes sad reading. This is not because the author busies himself with describing the miserable situation of the Soviet economy—a realistic description of the situation is a necessary prerequisite for successful change. The sadness engendered by the book lies rather in the feeling that the author has not really understood a crucial point: namely, the significance of prices to the workings of a market economy.

After having concluded that a lack of housing is the worst social problem in the Soviet Union, Aganbegyan maintains the view that house-rents as a proportion of income should be kept at about 3 per cent. How can the production of new houses and maintenance of old ones be encouraged when they must practically be given away? In countries with market economies, often 20–40 per cent of incomes will be spent on housing. But since a substantial part of this money will involve paying off mortgage loans, a person's own house will function as a savings bank for the individual household.

Further, Aganbegyan envisages a gradual introduction of the market economy, retaining state ownership of most of the means of production, and continued central control of the most important prices and quantities:

The socialist market is a regulated market in the sense that the prices for the most essential products will be set centrally, i.e., fuel, electricity, the most important raw materials, rolled steel machinery and some consumer goods. (Aganbegyan 1988: 119)

The interplay between supply and demand will thus not be given any room to develop. The role of prices as carriers of information to consumers and producers will not be heeded. And how could prices in Aganbegyan's world promote an appropriate co-ordination of all the decisions all the participants in the economy must make every day?

In the American magazine *Newsweek* (4 December 1989: 15), Soviet expert Allen Lynch, Assistant Director of Columbia's Harriman Institute, says:

I don't know of anyone in the Soviet Union who really understands what a market means, what its economic, social and political implications are—and that includes Abel Aganbegyan, who is one of the most advanced economists in the Soviet Union.

Changes in income, in preferences (i.e. the goods and services that consumers demand at any given time), and in technology will cause continual changes in the intersection between the supply curve and the demand curve. In a market economy changes in this point of intersection will result in changes in price and in quantity produced. A frequently changing state of equilibrium will thus have consequences for *what* is produced, *how much*, and at *what prices*. Perhaps most important of all, this movement in the direction of market-clearing will take place without the need for any omniscient planner telling producers what to do.

10.2. Prices and Relative Scarcity

When prices are determined via the interplay between supply and demand, they will express the relative scarcity of various resources. This favourable property of a market economy is called 'scarcity pricing'. And it is precisely this property that leads to existing resources being properly employed at any given time.

'Scarcity pricing' means that alternative ways of utilizing resources are measured against one another and compared. An investor considering shoe *or* shirt production will not need to make up his mind about what society needs most: he can merely calculate his own profit-making prospects and then choose the alternative that looks most promising.

When the capital-owner invests in such a way that he serves his own interests best, he will also serve society best. But—and this is an important qualification—this statement requires that all prices in the economy are right in terms of society's economic situation as a whole. As described in the preceding chapter, external effects, for example in connection with pollution, may result in this prerequisite not holding up.

Another and more problematical situation arises when the rate of unemployment is too high. In such a situation the price of labour (i.e. wage cost) is not the correct price in terms of the economy as a whole. In what follows we shall discuss the possibility of correcting imbalances due to incorrect pricing in connection with pollution and unemployment.

10.3. The Fight against Pollution

Pollution is an undesirable side-effect of productive activity and, sometimes, of consumption. Moreover, there is a clear limit to the total amount of pollution that a society can and should accept. Thus, pollution—or, more precisely, *permission to pollute*—will be regarded as an economic good. This means that there will be a 'correct' price for this good. Thus the question will be more pragmatic: How do we arrive at this price, and how shall we get participants to pay it?

If we take as a point of departure that we all—and through the authorities—are common owners of the clean environment, it will be up to the authorities to decide how they will organize the sale of 'the right to pollute'. Many people may certainly feel repugnance at the thought that, by paying for it, some people will be allowed to pollute. After closer consideration, perhaps they will realize the utopianism of the idea of living in a totally clean environment, where all pollution must cease no matter the cost. No one has ever managed to use energy without the emergence of undesirable polluting side-effects—be it poisonous gases in the atmosphere or horse manure in the gutter. Rather than arguing for or against pollution, the question is how much or how little pollution of various kinds a society should accept. Thus, we are talking neither of 0 nor 100, but of a suitable figure in between.

We could imagine two ways of setting a price on pollution: (1) impose a tax on polluters that is proportional to the amount emitted, or (2) determine the permitted amount of total emission and then make a market for the issue and sale of emission rights.

An advantage of the latter method, the sale of the right to pollute, would be that the authorities would gain control over the variable they were most interested in—i.e. the total extent of pollution. The sale of such rights could in principle be organized by means of *auctions*, where the individual bought the right to a defined quantity of emission for a quarter or half-a-year in the future. If experience indicated that the permitted amount of emission was too large, one could reduce the permitted amount at the next juncture, letting the market find its way to a new price, which in all probability would then be higher.

If one chose the alternative involving a tax, it would be reasonable to expect a trial-and-error period before the tax became sufficient to achieve the 'proper amount' of emission.

SUPPLEMENT 10.2. ECONOMISTS VERSUS THE PROPHETS OF REGULATION

In his well-written book, *Hard Heads, Soft Hearts: Tough-Minded Economics for a Just Society,* the American economist Alan S. Blinder (1987) concludes that economists have gained little acceptance thus far of their view of the best way to fight pollution. In an interview-survey of 63 environmentalists in the United States, there was not one who could explain why economists claimed that pollution could be combated more cheaply by using taxes than by exerting direct control. However, this lack of knowledge did not prevent many of those interviewed from being staunch opponents of taxes as an instrument in the fight for a cleaner environment.

As an example of the conflict between today's regulations and the economists' advice, Blinder uses as a point of departure a situation in which the authorities have decided that the amount of carbon dioxide must be reduced by 20 per cent. The prophets of regulation would then require all who released this toxic gas to reduce their emissions by 20 per cent. After all, that sounds just and reasonable. But would it be economically profitable for society as a whole?

No, says Blinder, and he gives an example from the real world. In a study in the large city of St Louis, it was found that the cost of reducing a certain type of emission from a paper mill by one ton was $4, whereas a corresponding reduction in emission from a brewery had a price-tag of $600. If both enterprises then were required to reduce emissions by one ton each, the cost to society would be $604. However, if the paper mill did the job entirely on its own and reduced its emissions by two tons, the cost would be a modest $8. In both cases the effect for society would be the same—the environment would be two tons cleaner. The 'just solution' would thus be $(604 − 8) = $596 too expensive.

If the authorities instead introduced a tax, for example of $100 per ton of pollution, the paper mill would voluntarily install cleansing equipment in order to avoid the tax. The 'profit' gained by this cleansing would be $(100 − 4) = $96 dollars per ton. The brewery, on the other hand, would continue to pollute, since paying a tax would be cheaper than installing a cleansing system.

Utilization of the market to fight pollution will have positive and dynamic effects. If one knows that pollution has a price—in the form of either a tax or the purchase of the right to pollute—there will be an incentive to develop new technology which results in smaller emissions.

Further, there will be an interest in reducing emissions irrespective of the level that exists initially. In the case of direct regulation, on the other hand, the enterprise will only have an incentive to cleanse as much as the regulations determine.

If, in the example from St Louis (see Supplement 10.2), we imagine that both enterprises were required to reduce their emissions from 1,000 tons to 800 tons, neither of them would see any point in a further reduction. If, on the other hand, the authorities were to introduce a tax of $100 per ton of pollution, the paper mill would reduce its emissions in the direction of zero (more precisely, to the level where a reduction in pollution of one more unit would result in cleansing costs of $100).

Making pollution cost something will motivate enterprises to behave in a societally appropriate manner. At the same time, there will be no necessity to give detailed regulations concerning choice of technology in production and cleansing equipment. The enterprise's own desire to obtain maximum profit, given a correct price on emissions, will see to it that the economic interests of society are looked after. Also in environmental policy it is important to recognize the fact that the market can be utilized to achieve societal goals.

It is in fact a general truth that behaviour which society wishes to curtail becomes the subject of taxation; in this context we have mentioned taxes on tobacco earlier in this book (see Supplement 2.1), and a similar argument applies to alcohol. In the same way as consumers adjust to taxes on consumer goods, producers will adjust to taxes on pollution. In both cases the intention is to change the prices a free market generates without taking away or interfering with the market mechanism. Thus, consumers and producers will retain their free choices, but this time in relation to market prices, which take into account the consequences for the economy as a whole.

In connection with pollution, use of taxes or the sale of the right to pollute can be utilized to ensure a correct price for the environment. With the right price here, the decentralized market can continue to do its job as co-ordinator of the individual economic agents' decisions. There will be less need for a command system where everything is controlled by regulation-happy bureaucrats.

10.4. Unemployment

Our second example of erroneous pricing in the economy, namely, the fact that the price of labour may seem too high when there is high unemployment, raises even greater and at times insoluble problems.

As a starting-point, we must realize that a certain degree of unemployment is compatible with full employment—even though this may sound strange. In a market economy, the fact is that there will always be people who are just entering the labour market, while others will have resigned from their jobs and be searching for better ones. It is natural for people to move between jobs and places and to be unemployed during the transition period.

This *natural* (or normal) *unemployment* rate varies from country to country. In the United States, where people change jobs relatively frequently, it is normal to have an unemployment rate of between 5 and 7 per cent, as we mentioned in Chapter 9. Corresponding figures for Japan, Sweden, and Norway are less than half of this, i.e. around 2–3 per cent. The natural rate of unemployment for most other market economies lies between 2 and 7 per cent.

Unemployment in excess of the norm may be due to various factors originating on the supply side or the demand side of the economy. Keynes thought that insufficient aggregate demand often played the most important part. A lack of desire to invest in the business sector or a reduction in export revenues may lie behind a decrease in aggregate demand for goods and services. And when enterprises' sales go down, the need for labour will also be reduced.

However, joblessness can also originate on the supply side of the economy. If employees are well organized in labour unions, they can force through significant wage increases. When labour becomes unacceptably expensive, management in enterprises will choose to utilize less labour and more machinery in the production of a given quantity of goods. Moreover, goods will become more expensive when wages rise. For enterprises that compete with foreign producers, this may mean that they will lose market shares both at home and abroad. Thus, too high wages may result in total domestic production falling. This will also lead to a reduced demand for labour.

Under conditions characterized by full employment, production will go down in enterprises that lose labour and will rise in enterprises where the work-force expands. For society as a whole, the individual's change of job will involve an opportunity cost, namely, a fall in production in the enterprise she leaves.

In a situation with unemployment in excess of the natural rate, the picture will be different. When a person who is unemployed gets a job, this will not lead to a loss in alternative production. In an economy where there is significant unemployment, we can thus say that the cost to society of employing one more person will be practically zero.

In order to understand this reasoning better, it may help to remember that the price for society of any resource will be the best alternative forgone when the resource is put to concrete use in production. When the resource—in this case, labour—stands idle, the production that is sacrificed by employing an additional worker will be small. The unemployed person might perhaps spend her time making clothes or redecorating her flat. When she is in full-time employment she will have less time for such activities. And the production lost by society because she starts working will be less clothes-making and redecorating at home. However, in its calculations the enterprise cannot set the price of new labour at zero or close to zero. It will, after all, have to pay the going wage. In such a situation there will therefore be a difference between the price of labour from the point of view of the enterprise and the same thing viewed by society. In other words, the picture will be characterized by a *market imperfection*. The way should thus be paved for intervention by the authorities, in order to ensure that the enterprises will be confronted by the correct price of labour in terms of the total economy.

10.5. Ways to Handle Unemployment

In a situation of unemployment we could in principle imagine three options for the authorities. The first option would be to reduce wage costs for the enterprises. This could take place by means of direct subsidization of labour. For example, for several years Norwegian ready-to-wear clothing and textiles enterprises received a public sector subsidy of Nkr 2 per completed working hour. This constituted an estimated 5 per cent of the enterprise's wage costs.

There are a number of problems linked to such subsidization. Actually, the authorities were only supposed to subsidize new jobs. In practice, it is hard to prevent existing jobs from also being subsidized. Moreover, it is difficult to determine which business sectors or enterprises 'deserve' to be subsidized. The awareness of the possibility of subsidization will encourage enterprise managers and industrial organizations to take up 'lobbying'. Instead of working on the economic challenges facing the industry or enterprise, time and energy will be spent on influencing politicians and bureaucrats; the purpose is to convince the public authorities that *their* branch or firm should be given the benefit of subsidized labour. Using our earlier terminology, we could say that the private business sector makes efforts to 'soften' the 'hard' budget constraint to some degree.

Finally, there is little point in maintaining life artificially through various subsidization schemes in business sectors that will eventually lose out to competition from foreign producers. As regards Norwegian ready-made clothing and textiles producers, the subsidies only helped to prolong the phasing-out of the industry, not to prevent it.

A second way in which the authorities can attempt to alleviate unemployment is to stimulate aggregate demand in the economy. We will remember from the preceding chapter that this was what Keynes recommended in the 1930s: that the state should increase its own demand for goods and services and in this way see to it that more people entered into paid employment. This can be done by means of increased public expenditure or reduced direct and indirect taxes, as we discussed in Chapter 5. In many situations this will be a beneficial and correct policy. However, at times it may also be problematical. Let us focus on a couple of factors.

First, it will be difficult for the authorities to know when and how strongly the economy should be stimulated. Experience gained from attempts at 'fine-tuning' market economies shows that sometimes the instability in the economy has been reinforced. Measures aimed at stimulating the economy were implemented when the economy had passed its lowest point and was on its way up. At times the measures used by the authorities involved too strong a stimulus. The result was growing inflation in the 1970s (see Supplement 9.2).

Second, an increase in public expenditure will necessitate a certain increase in public revenues. However, increased direct and indirect taxes, which are the authorities' most important sources of revenue, will create their own problems. High taxation will create incentives for tax evasion and unregistered economic activity (see Supplement 6.2). And, when they are not justified on the basis of the desire to correct for external effects such as pollution, the various taxes may in themselves lead to price distortions that can have harmful effects on the economic system—for example, by reducing incentives to work and save. This means that taxation in itself can create *market imperfections*. In such a case one type of market imperfection, i.e. too high a wage level with correspondingly high unemployment, will be replaced by another one, i.e. erroneous prices due to excessive taxes.

A third way in which the authorities can meet unemployment in excess of the natural rate is to remain passive: that is to say, they can refrain from intervening with regulations (among other things), since households and enterprises have learned how to 'fend off' such measures. Well aware of the fact that economic activity fluctuates, and that various forms of active control of aggregate demand can be extremely difficult, throughout the 1980s the authorities in most Western countries substantially reduced their confidence in politico-economic measures in this area. Instead, there was an increasing focus on the supply side. That has meant increased concern about what we have allocated much space to in this book: namely, how to ensure optimum utilization of existing resources in the long term.

One example of the fact that the supply side has been paid greater attention is all the work within the European Community (EC) aimed at increased competition within the various business sectors and between the various countries (see Supplement 5.4). Another is the negotiations within GATT (General Agreement on Tariffs and Trade), where the objective is the further dismantling of trade barriers, particularly for agricultural products and trade in services (e.g. transport and insurance). A third example is the increasing interest in government investments in further education at all levels (life-long learning).

To summarize, in connection with pollution, the authorities can demand a price for the right to pollute and in that way use the market to resolve market failures. In the case of persistent unemployment far in excess of the norm, most economists would recommend an active contribution on the part of the authorities. However, no one could give any guarantee that everyone who wanted work at any given time would be sure of getting a suitable job. The 'hard' budget constraint under which the participants in a market economy work will guarantee an organizationally adaptable economy. The other side of the coin will be periods characterized by too high unemployment.

10.6. Individual Freedom and Economic Uncertainty

The reason why this chapter is entitled 'Why the Market Nevertheless is Useful' is linked not least to the final point we shall take up here. Individual freedom, which is a prerequisite of democracy, can hardly be realized under the system of economic planning, which is the only known alternative economic system.

Although the intentions of the architects of a command or planned economy may be the best, including a rational utilization of resources and a more even distribution of income, experience shows that the means that are utilized are always misused. In a planned economy, where virtually all economic activity is directed from the centre, those who make the decisions there will gain great personal power and influence over others. This will open the way for arbitrariness. The individual's economic situation will depend upon the degree to which he can curry favour with the powers that be.

In a centrally controlled command economy there is no room for private initiative, with its accompanying possibilities and risks. And when the state stands as the owner of all the means of production, and thus the sole employer, the individual loses the possibility of choosing whom to work for. The absence of choice—whether to start one's own economic undertaking or to change employer or job or residence—is a special feature of the planned economy. And when the possibility of choice does not exist, one loses much of one's freedom.

In a market economy based on competition and private ownership, one individual's chances of influencing other people's incomes and positions will be far smaller. When ownership is divided among many, the individual will be free to choose where he wants to seek work. If he dislikes his job, he will be free to seek a job with another employer. Perhaps he may have to accept lower wages; however, he will retain the right to make his own choice.

The more enterprising individual, who has capital himself or can persuade others to place capital at his disposal, either as loans or as equity capital, will have the right to start his own business. In such a case he will assess the risk of starting a new undertaking against the prospects of profits. Should he elect to resign from a safe job to start his own undertaking, he will be using his freedom, and should be well aware of the price he must pay: namely, greater uncertainty concerning his future private economic situation.

When one single comprehensive plan forms the basis for all economic decision-making in society, the individual's economic freedom is substantially reduced. Rather than seeing the satisfaction of his material needs and desires as the *goal* of economic activity, the individual is reduced to being a *means* of achieving the goals of planning. The result can be read in the composition of production in planned economies: enormous investments directed by the goals of planning, and stagnation in the production of consumer goods. But the latter situation—i.e. production of more and better goods and services for private consumption now and in the future—should in fact be the objective of economic activity in society.

Whereas production exists for the benefit of the individual in a market economy, it seems as if the individual exists for the benefit of production in a planned economy. Planning has also proved to be incompatible with openness and the free flow of ideas and information. Hayek expresses this thought succinctly in his book *The Road to Serfdom*: 'The probable effect on the people's loyalty to the system becomes the only criterion for deciding whether a particular piece of information is to be published or suppressed.' (1944: 160).

Despite Gorbachev's openness (or *glasnost*), the way in which the authorities handled the Chernobyl accident in April 1986 illustrates Hayek's point. Thousands of Soviet citizens fled to railway stations in an attempt to get away from the radioactivity resulting from the accident—but the only information they could get about what was happening was from foreign radio broadcasts.

The freedom made available by a decentralized market economy is linked to a certain degree of economic uncertainty. This is due to the fact that the economic system is not a deterministic system. Its inherent uncertainty is something society cannot free itself from completely.

We have previously mentioned that good and bad luck may play a part in determining how well the individual does economically. A person who has chosen an occupation which in time is made redundant by economic developments may have good grounds for feeling cheated. In times past, however, as a result of the transition to steam, sailmakers experienced a fall in demand for their skills, as did whalers when whales were declared a protected species. The coachman with his horse and wagon was perhaps somewhat better off; if he managed to make the transition to the motor car, he could continue his production of transport services, but now with new technology and perhaps with a new employer.

Since the end of the 1970s typographers in all Western countries have been forced by circumstances to adopt new methods or change jobs. Modern computerized equipment is in the process of making an old craft superfluous. When some typographers lose their jobs and others must accept lower wages, this gives important signals concerning the occupational choices of future generations. The market as a *system* demonstrates its flexibility, but also its 'unfairness', since good and bad luck will always play a certain part.

If, in this situation, society were to guarantee typographers their jobs and wages for the rest of their lives, this group would gain a privilege above the rest of the work-force. At the next juncture, others who were in a correspondingly unfortunate situation could reasonably make the same demands. The economy's adaptability would gradually be eroded. And the security that each group achieved would result in increased insecurity for the remainder. The cake to be shared would become smaller, without there being any guarantee that the individual slices would be more equal in size.

To summarize, the individual will have far greater freedom of choice in a market economy than in a planned economy. This will apply both to employees and consumers. This economic freedom is of fundamental importance to other types of freedom, such as freedom of information and freedom of speech.

However, economic freedom does not involve the absence of economic worries. Perhaps the greatest problem with a market economy will be periods when unemployment is in excess of the natural rate. As a result of this, among other things, the electorate in a democratic market economy will demand social welfare schemes which guarantee everyone a minimum standard of living. What this minimum should be will always be a bone of contention. Obviously, it will change over time; that is to say, it will rise in line with the general level of production in society. The material standard of living of a basic-level pensioner in the Scandinavian countries today is considerably higher than the standard enjoyed by many full-time workers at the beginning of this century. And it is also higher than what many workers in Central and Eastern Europe must be satisfied with today.

In addition to the fact that morality and a sense of justice would indicate that no one should have to experience hunger or deprivation, the security of a set minimum will be a necessary prerequisite for many who wish to take the risks entailed by a market economy. In order for the enterprising man or woman to take the chance of starting a new undertaking, he or she must be sure of surviving if the project should be a failure.

10.7. Summary

A perfect market economy cannot be found. It does not exist. A situation where all prices are right at any given time, thus being the carriers of correct information, will never come into being. Market imperfections of various kinds are something a market economy will always have to struggle with. These imperfections must be measured against the imperfections that direct intervention by the authorities will often lead to. Before the authorities choose to correct for market imperfections, they should therefore be quite certain that the regulations that are introduced do not make matters worse.

Moreover, the desire for total economic security for all will never be possible to realize completely. The loss of individual freedom and of economic efficiency would be too great for this to be possible. What the British statesman Winston S. Churchill once said about democracy could be paraphrased to apply to the market economy:

Many forms of government have been tried and will be tried in this world of sin and woe. No one pretends that democracy is perfect or all-wise. Indeed it has been said that democracy is the worse government except all those other forms that have been tried from time to time. (House of Commons, 11 November 1947)

PART III

Public Finance, Money, Capital, Labour, and International Trade

In Part II we gave a general introduction to the workings of the market. In addition, it was made clear why and how the market economy generally is superior to central planning as a co-ordinating mechanism in the economic system.

In Part III we shall look more carefully at the economic role of government and of money in modern market economies. We shall also consider the structure and functioning of the labour market, the capital market, and the international market for goods and services.

We will again meet terms such as 'supply' and 'demand', 'scarcity pricing', 'market-clearing', and 'externalities' or 'external effects'. In addition there will be new ones, such as 'public goods', 'competitiveness', 'return on investment', and 'comparative advantage'.

A thorough understanding of public finance and money and of how the three markets mentioned function will begin to enable readers to form their *own opinions* on economic policy. They will see that some simple questions do not always have simple answers. What should the role of government be in a market economy? Should government be 'big' or 'small'? How should the government raise necessary revenue? Sensible people may take differing views on this.

To take another example, even if there is general agreement that free trade with foreign countries is important for economic efficiency, there may be disagreement regarding how *rapidly* existing trade barriers should be removed. Too rapid a removal might result in enterprises that were actually viable disappearing into bankruptcy. Too long a phasing-out period will result in inefficient utilization of resources.

Economics is basically a matter of trade-offs. 'You can't have your cake and eat it,' as the English proverb says.

11

The Political Economy of Public Finance

> Taxes are what we pay for a civilized society.
>
> Oliver W. Holmes

THE transition from central planning to a market economy entails a drastically reduced scope of government involvement in economic affairs as well as in other spheres of life. Each individual household will need to strive to attain the greatest possible material welfare, given its limited resources. And each firm will have to be guided by its desire to maximize profits.

However, the fair and free play of market forces is not an end in itself. Rather, it is a means to foster economic efficiency and steadily improve living standards of the people. Also, the market system seems more compatible with individual liberty and political pluralism than the alternative of a command economy.

11.1. The Economic Role of Government

This having been said, in a market economy the government is not relieved of economic responsibilities. All market economies are mixed economies—in the sense that, while most economic decisions are made without government interference, some decisions *are* entrusted to the government for various reasons. We have already discussed, in Chapter 7, three basic tasks of the government in a market system: (1) to ensure stable rules that protect private ownership, (2) to ensure effective competition and free exchange of goods and services, and (3) to provide a legal framework for bankruptcy and industrial reorganization. These basic preconditions are essential for a market economy to function properly.

Beyond this, the government has a potentially important and useful economic role to play for other reasons, some of which we discussed in the preceding chapter. Markets are not infallible. When they fail, the government can sometimes help to put them in order. We reviewed two important examples of market imperfections: environmental pollution and unemployment. However, not only markets can fail, but also governments. Therefore, it is important always to make sure that market failures that are alleviated or eliminated by the government are not replaced by government failures of similar or even greater severity. In economic policy as in medical practice, the cure must never be worse than the disease.

The extent of government involvement in economic and social affairs has been at the heart of the controversy concerning the comparative merits of central planning versus a liberal market economy, and of political controversy in general, at least since the days of Karl Marx.

Two or three decades ago it was a widely held political view in Western Europe that the government had an important leadership role to play in industry and in various kinds of social services where private enterprise was considered impracticable or undesirable. Examples include banking, education, health service, insurance, transport communications, and broadcasting. This opinion was widespread even among those who did not favour general nationalization of industry.

In the light of experience, this view has gradually changed. This is not surprising; markets sometimes fail, it is true, but so do planning bureaucracies and political authorities. Nowadays, therefore, it is more commonly believed that private enterprise and ownership should be the general rule, not least in those fields where extensive government involvement has been tried with mixed results. The wave of privatization of government-owned enterprises all over Western Europe and elsewhere in recent years has been the result of this change of general public opinion.

At the same time, new questions have been raised about the traditional three-pronged role of the government in economic affairs in market economies: first, to promote macroeconomic stability, including full employment without inflation; second, to ensure a reasonably just distribution of income and wealth; and, third, to foster an efficient allocation of productive resources among alternative uses. The doubts that have awakened stem in part from the realization that the governments of many countries have not succeeded very well in promoting macroeconomic stability in the turbulence of world economic events over the last twenty years. In some cases, government attempts to stabilize economic activity actually seem to have had destabilizing effects.

Besides, there is naturally no general consensus in society about what is 'fair' and 'just' in economic life—or elsewhere in human affairs for that matter. Questions of fairness and of social justice are essentially ethical and philosophical in nature, and therefore lie outside the purview of pure economics. The 'right' extent of government involvement in economic and social affairs, therefore, is not solely an economic question that, like most of the economic issues addressed in this book, can be settled by appealing to scientific arguments and experience. The suitable extent of government involvement is also an inherently political question.

This brings us to an important distinction that must be made throughout economics, especially in public economics: namely, the distinction between positive and normative economics.

Positive economics deals with the way things *are*. The statement that a decrease in the price of shoes will increase the demand for shoes, other things being equal, is a positive statement. It can be proved right or wrong by appealing to empirical evidence or experience. (There is, incidentally, overwhelming evidence that a decrease in price increases demand in the real world, other things being equal.)

Normative economics, on the other hand, is about the way things *ought to be*. Thus, the statement that, for some purpose, the price of shoes or prices in general should be made to fall, presumably through government action (how else?), is a normative statement. It cannot be proved right or wrong. The answers to normative questions cannot be settled solely and unequivocally by scientific arguments or evidence because the answers to such questions depend on value judgements. And value judgements differ from one individual to another. Many questions in public economics are of this nature; hence the title of this chapter.

But this is not to say that positive economics cannot be brought to bear on problems that arise in demarcating the economic role of the government in modern market economies. An example will illustrate the point. The extent to which the government should endeavour to make the distribution of income in society more even—by providing an economic safety net for disadvantaged social groups, among other things—is essentially a normative question. None the less, an informed answer to this question would depend in part on positive judgements concerning, for example, the effects of increased government interference and of increased income equality on other things society aims for, such as economic growth. There are indications in recent research that too much government interference as well as too much income inequality may be harmful to growth. If so, prudent governments should try, through positive economics, to obtain a clear picture of what combinations of income distribution and economic growth are feasible. With such information at hand, a decision based on ethical considerations must be made.

11.2. Externalities

The new sceptical attitudes towards government involvement in economic life do not imply that the government has no useful economic role to play in modern market societies beyond the basic tasks reviewed at the beginning of this chapter. Consider a person who holds the view that the government should not try to promote macroeconomic stability or influence the distribution of income and wealth, but should rather accept the outcome produced by the interplay of free market forces in this respect: this person may nevertheless consider it legitimate for the government to attempt to foster an efficient allocation of factors of production, within reasonable limits.

This has to do with the fact that some kinds of production and consumption affect not only the producers and consumers themselves, but others as well. As long as *external effects* of this kind are present, the government has a legitimate and justifiable reason to intervene. This is the main reason why the government continues to play an important role in health care, education, arts, and science in all modern market economies. On the other hand, banks, insurance companies, airlines, railways, and radio and television stations have gradually been transferred to private ownership in many countries in recent years. External effects are quite limited in these latter areas.

As an example of negative externalities in consumption, consider smoking. Now it is common knowledge that smoking is dangerous to your health, both directly and indirectly. It is not your private affair whether you smoke in the presence of others, since they too are affected. Even when you smoke alone it is not solely a private affair, because if you develop heart disease or lung cancer from your habit you impose a burden on your fellow citizens who pay for health services or health insurance or both. Therefore, it is now generally considered natural for the government to try to combat smoking, primarily by levying taxes on tobacco. In fact, the European Community aims to eliminate the smoking of cigarettes from the Continent before the turn of the century. Be that as it may, it is generally considered reasonable to levy taxes or fees to restrain individual behaviour that is harmful to others. The idea is, in other words, to combat negative externalities by tax or fee. Fee is actually a better word than tax in this context, because smokers are being charged for using and polluting common property, namely, the air that we all breathe. This is known as the *polluter-pay principle*.

Traffic regulations are another case in point. It is not our business alone whether we drive on the right or left side of the road, or how fast. This is why the government or some public authority sets traffic regulations. And, when you think about it, it is not solely a private affair whether we drive a car at all or not because, in heavy traffic, drivers impose delays on one another. This is why the government levies road tolls, parking fees, and so forth. And this is why it would make sense to impose taxes or fees on traffic on congested roads in general in order to spread the traffic and reduce delays—and to create scope for reducing other less efficient forms of taxation. We shall return to this issue later in this chapter.

11.3. Health Care and Education

On similar grounds, the government may rightly want to encourage individual behaviour that has beneficial effects on others. It can, for instance, be argued that *positive externalities* stem from good public health in general, in the sense that each individual in society benefits from the good physical and mental condition of others. If so, it can be argued moreover that the government should provide or guarantee, below the free-market price, health services beyond what would be forthcoming in a free market.

But this is not to say that hospitals must necessarily be owned and operated by the government rather than privately. It is quite possible that the government could provide better health service to the public at less expense by fostering private hospitals and health insurance. In some countries, notably the United States, hospitals are privately owned and operated in general, even though the US government provides free or subsidized health care for old people and the poor. But because the positive externalities involved are not obvious, the role of the government in general health care continues to be controversial in America and also in Europe.

Preventive medical care aimed at limiting the spread of contagious diseases is clearly a legitimate governmental concern, however, because of the negative externalities involved. The cure or prevention of a contagious disease is worth more to society than to the individual carrying the disease, so the government has reason to intervene.

Education is another good example. Not only the educated benefit from living in an enlightened society. This is why education is compulsory and why the government operates public schools or supports private schools at all levels throughout the world. Some people argue that increased competition, decentralization and privatization, and a freer play of market forces in general could make schools more effective by securing more and better education at less public expense. There would nevertheless be a justifiable reason for government support of education of many kinds. Without government support, too little education would be supplied in free markets because the direct benefit of education to each parent or student is less than the benefit that accrues from education to society as a whole.

11.4. Arts and Science

Similar reasoning applies to arts and science. The results of scientific research—the works of Louis Pasteur and Wilhelm Conrad Röntgen, and of Adam Smith and John Maynard Keynes, to name but four prominent examples—benefit many more than the scientists themselves and their collaborators and families. This is why the government supports scientific research all over the world, either directly by grants from public funds or indirectly by providing tax concessions to individuals and firms that support scientific research.

The problem here is that scientific work does not necessarily pay from the point of view of individuals and firms, even though it can be enormously profitable from the vantage point of society as a whole. Since the benefits of scientific research to society exceed the private gain, without public support the private sector would tend to neglect science. Consequently, society would reap less scientific knowledge than desired if the government did not bridge the gap.

This explains why universities are owned and operated, or at least supported financially, by governments all over the world. Private universities in the United States and elsewhere would encounter great financial difficulties if individuals and firms were not given encouragement, through tax concessions and other means, to support them.

Scientific research is a *public good* to a considerable extent. By this is meant that my acquisition of the good does not reduce your access to it. Thus, the fruits of scientific research are generally shared by all. If one benefits, most or all do, because scientific knowledge travels easily between people and places.

A similar argument applies to the arts. The government supports arts and culture all over the world on the grounds that music, literature, painting, etc., are widely enjoyed not only by those who purchase tickets to cultural events and by the artists themselves, but also by the community at large. In the minds of most people, the quality of life is greater in a community rich in arts and culture than in an artistic and cultural void. By implication, free markets would supply too little art and too little culture from the point of view of society as a whole.

But the arts are not only a public good, but also a *merit good*. By this is meant that art and cultural activities are considered valuable or desirable *as such*, independently of the external benefits involved. For this reason, too, most governments consider arts and culture deserving of public financial support. By the same token, it seems probable that most governments would consider it right to tax smoking even if smokers harmed nobody but themselves.

This list of examples can be extended. Agriculture is comparable with the arts, for instance, in that it confers benefits not only on farmers and their families but also on city dwellers who appreciate the beauty of the countryside. The positive externalities that stem from farming constitute legitimate grounds for government support of agriculture, within reasonable limits.

Before concluding this discussion, we should emphasize that the externalities argument for government intervention in certain areas of the economy that would be undernourished otherwise must not be taken too far. A prudent government does not subsidize every pursuit that may impart a positive externality, nor does it tax every activity that creates a negative one. There are two important reasons for caution here. First, too much government interference may have detrimental effects on the economy if it leads to an excessive expansion of the public sector at the expense of more efficient private economic activities (see Supplement 11.1). Second, government expenditures must be financed somehow, and the financing, in its turn, has an important bearing on the structure and functioning of the market system. This brings us to our next topic: taxes.

SUPPLEMENT 11.1. THE EXPANSION OF THE PUBLIC SECTOR IN MARKET ECONOMIES

A substantial increase in government economic activity has been a salient feature of most market economies over the last twenty or thirty years. A few figures will illustrate the point. The share of total government outlays as a percentage of gross national product in the United States rose from 27% in 1960 to 36% in 1989. In Japan and Sweden, this share almost doubled over the same period; in Japan the share of total government expenditure of GNP grew from 17% to 33% between 1960 and 1989, and in Sweden it grew from 31% to 60% in the same period. Japan and Sweden have similar GNP per capita, but Swedes choose to channel almost twice as much of their national income through the public sector as the Japanese.

Why has the public sector expanded so rapidly in these countries and in virtually all other market economies over the years? The development of the United States towards a modern welfare state that began in the 1960s is an important part of the explanation there. Similarly, ambitious welfare policies aimed at equalizing the distribution of income and wealth on an unprecedented scale have provided impetus to the expansion of the public sector in Sweden. Similar country-specific explanations can be given for other nations.

But this is not everything. There is an automatic and built-in tendency in democratic societies towards the expansion of the public sector. This tendency stems from the uneven distribution of the benefits and costs of government spending. The beneficiaries of each new government programme are usually few compared with the large mass of taxpayers who ultimately must foot the bill. Various vocal groups and organizations pressurize the government to build a bridge here, subsidize a factory there, and so forth. Frequently, a lot is at stake for each of those who benefit from increased public spending, while the interests of taxpayers are diffuse. Government subsidies to tobacco farmers in the United States mean a lot to each farmer in Virginia, for example, but they cost each taxpayer in the country as a whole a fairly small sum. Therefore, from the point of view of each individual, it makes better sense to join groups that fight for greater public spending than to resist higher taxes. The result is a steadily expanding public sector.

However, the sky is not the limit here. If the expansion of the public sector is considered to have gone too far, the general public or the government in a democratic country will almost surely take measures to break or reverse the trend. Sweden is a case in point: there, the share of total government spending in GNP was reduced from a peak of 66% in 1982 to 60% in 1989.

Source: Statistics taken from *OECD Historical Statistics 1960–89*, Paris, 1991

11.5. The Inflation Tax

The government has three principal ways of generating the funds necessary to finance its expenditure: printing money, levying taxes, and borrowing.

Printing money seems tempting. After all, the government has full control over the mint, or money printing press, which is exclusively public property everywhere in the world. This method is not as far-fetched as it may sound. In fact, it is fairly commonly used, especially in developing countries and in southern Europe and Iceland. In these countries other sources of financing are sometimes difficult to come by.

The problem with printing money to finance government expenditures, however, is that it causes inflation. We shall study this more closely in the next chapter. Here it suffices to note that an important part of the reason why developing countries generally have more inflation than developed countries is that the governments of many developing countries finance relatively more of their spending by printing money. This is also the main reason why inflation has skyrocketed in the Soviet Union since the mid-1980s: government spending rose, tax receipts fell as production plummeted, and the government in Moscow bridged the resulting budget deficit by printing money on a massive scale. One more example comes from the United States and has already been mentioned in Chapter 9. High rates of inflation in the United States in the early 1970s were triggered partly by an expansion of military expenditure that was financed in part by printing money. Other examples abound.

The fact that government spending can be financed by printing money implies that inflation is tantamount to a tax. To see this, consider the dilemma of a government that is determined to build a monument but is reluctant to raise taxes to pay for it, out of fear that the general public would not approve. To bypass the public, the government can order the central bank to print money to cover the construction costs. The monument rises without taxes having been raised. So who footed the bill?

The public did. How? The purchasing power of wages and pensions, and of money and other financial assets held by the public, declined as a result of the inflation caused by the expansion of the money supply. The public paid for the monument—not directly, through higher taxes, but indirectly, through reduced purchasing power and a lower standard of living.

The *inflation tax* collected clandestinely in this manner has been an important source of revenue for the governments of many high-inflation countries over the years. Some governments have actually been reluctant to reduce inflation partly because they have become dependent on inflation tax receipts.

11.6. Taxation without Inflation

To finance public expenditure without recourse to monetary expansion and inflation, the government must either raise taxes or borrow (at home or abroad), or both. In most market economies the bulk of government expenditure is financed through the collection of various taxes and fees. If the tax receipts do not suffice to cover expenditures, the government must borrow from the public or abroad to bridge the gap.

Consider taxes first. They take several forms in modern market economies. The most important are *income taxes*, which are levied on the individual on incomes from wages, pensions, and various assets (such as interests earned on bank deposits); *excise, sales, or value added taxes*, which are levied essentially on expenditures on goods and services; *profit taxes*, levied on companies; and *tariff duties*, which are levied on imports. Each of these types of tax has disadvantages. The basic problem with all of them is that they have unintended side-effects that tend to distort economic decision-making and reduce efficiency.

For example, income taxes discourage people from working and producing. Excise taxes, by sparing savings, fall relatively heavily on low-income families that spend most or all of their incomes on necessities and save almost nothing. Taxes on profits can induce enterprises to overinvest in inventories and machinery; such investments will increase the amount of depreciation (more on this concept in Chapter 17), and reduce the level of profit. This implies also a reduction in taxes levied on the enterprise. Tariffs discourage imports from abroad, and hence reduce the efficiency gains from foreign trade, as we shall discuss in greater detail in Chapter 15. The inflation tax, covered in the preceding section, is unfair in the eyes of most people. This tax falls most heavily on those whose wages, pensions, and savings are fixed in money terms; their fixed nominal income loses value when prices rise. The inflation tax is also inefficient because of the problems associated with inflation; more on those in the next chapter.

Not all forms of revenue collection are inefficient or unfair, however. Pollution fees are a good example. Discouraging activities that damage the natural environment (driving or deforestation, for instance) by charging a fee serves to clean and preserve the environment. The side-effects of such 'taxation' are generally favourable. Rather than reducing incentives to work and produce like income taxes do, pollution fees primarily encourage households and firms to keep the environment clean for the benefit of all. Pollution fees are not 'taxes' in the ordinary sense of the word: they are better described as fees for using and polluting a common property, the environment.

The government has a responsibility to ensure that the tax system is as efficient and as fair as possible. When efficiency and fairness are at odds, a reasonable balance between the two must be reached. As a general rule, inefficient taxes should be replaced by more efficient ones whenever possible, other things being equal.

Fairness is more difficult to judge objectively. If the government's aim is to make the distribution of income more equal, as is almost always the case, it may be considered appropriate to apply a higher income tax rate on rich people than on poor (*progressive income tax*). The income tax paid by the poor may actually be negative, in the sense that the poor receive income support from the government in inverse relation to their income (*negative income tax*). This being the case, one ought to ensure that the support they receive does not reduce their incentive to work.

In general, tax burdens should be linked in some way to the benefits taxpayers receive from the government, and expect to receive in the future (e.g. in terms of publicly paid pensions), as well as to their ability to pay. However, the right balance between direct and indirect taxation (that is, the choice between using income or expenditure as the basis for taxation) and the right degree of progressivity of the income tax depend in a complicated way on the circumstances of each country. These things cannot be determined conclusively once and for all.

11.7. Deficits and Debts

We now turn to the third and last method that the government can use to finance its operations. If the tax revenues of the government fall short of its expenditures, and if resort to the printing press is out of the question because inflation is to be avoided, the government must borrow from the public, at home or abroad. Let us return to the example of the monument discussed above. Government borrowing usually takes the form of issuing and selling interest-bearing government bonds to the public. The money thus collected can finance the construction of the monument without taxes being raised and without any new money being printed. What happens now? The government receives money from the public, and uses it to pay for the construction. So the monument rises. The public receives bonds from the government in return and, having just bought the bonds, has correspondingly less money to spend on other things or to save.

This is not the end of the story, however. The government must pay interest on the bonds and, ultimately, it must also repay the principal (unless the bonds are of infinite maturity). By running a bond-financed budget deficit, the government has incurred a debt which must be serviced. Interest payments on the public debt can easily become a substantial expenditure item in the government's budget.

If the national debt is all owed to domestic residents, servicing the debt only entails a redistribution of income within the country; all the interest and amortization payments stay at home. In this case, the government levies taxes in the usual way, or borrows, in order to meet its domestic debt obligations and other expenditure needs.

If, on the other hand, the government has sold its bonds abroad, the subsequent payments of interest and principal on the foreign debt thus incurred will flow out of the country. Consequently, if the foreign borrowing has not been used to increase the income potential of the country by investing in profitable ventures, less resources will be available to satisfy domestic needs. This can be a very serious problem; some countries in the developing world have to part with a third or more of their export earnings every year to meet their foreign debt obligations, incurred largely by reckless government spending in the past (see Supplement 11.2).

SUPPLEMENT 11.2. DEBT CRISES

Inflation is not the only hazard involved in government spending in excess of tax collection and domestic borrowing. Government budget deficits financed by foreign borrowing may invite double jeopardy: inflation and a debt crisis.

Many of the most heavily indebted countries of the world are among those with the highest rates of inflation. This is not a coincidence. Why? Foreign borrowing is more inflationary than domestic borrowing in general, other things being equal. Borrowing from domestic residents reduces their purchasing power, and thus dampens the inflationary effect of the expansion of bond-financed government spending. Borrowing abroad, on the other hand, has no such dampening effect. In this case, the increase in public expenditure is not offset by a decrease in private expenditure. Therefore, government borrowing abroad has a greater tendency to raise the price level than borrowing at home.

But that is not all. Deficit spending and debts tend to accumulate by themselves. Interest payments feed back on the budget deficit, and hence necessitate further borrowing in the absence of tax increases or spending cuts. In recent years, many countries have found themselves in a vicious circle of this nature, their foreign debts spiralling out of control. More than a dozen countries in the Third World presently have accumulated foreign debts that exceed their annual gross national product. This means that it would take the people of these countries more than a year to earn enough to pay back their debts; meanwhile they would have no income left over for other things. These nations have to spend a substantial portion of their export earnings to service their debts.

Argentina, Brazil, and Mexico spent about one-third of their export revenues on debt service in 1988, for example, and Hungary almost one-fourth. Debt service of this magnitude can be a heavy burden to carry, especially if the proceeds of the foreign loans were not invested in projects that increased the earning potential of the country and hence its ability to pay back.

But why pay back? Apart from the moral aspect of not honouring one's debt obligations, there is a practical problem. A sovereign nation going back on its foreign debt without the consent of its lenders would almost surely be excluded from world financial markets for a long time. Not only would it not receive any more fresh loans, but, by reference to international law, its assets abroad could be confiscated, its aircraft could be stranded abroad, and so on. In practice, however, when the amount of a country's foreign debt gets out of hand, it is renegotiated rather than repaid. Sovereign nations do not go bankrupt in the real world. But this is not to say that they can live far beyond their means for ever. Sooner or later, they must make ends meet.

12

The Role of Money

Money . . . is not the wheels of trade: it is the oil which renders the
motion of the wheels smooth and easy.

David Hume

MONEY is the generally accepted means of payment. Throughout history
various goods have served as money. In an historical perspective, human
beings have probably been most accustomed to using various forms of
metal as money. Coins of copper, silver, and gold date from the earliest
times. They are described in the Old Testament.

12.1. Metal Coins and Paper Money

The advantages of utilizing metal coins of various kinds are many. It is
relatively easy to divide them into standardized sizes; they are durable;
and perhaps most important, they have an inherent value. By this we
mean that gold coins, for example, can be melted and used to make rings,
bracelets, or other ornaments.

In the state of Virginia in the United States, tobacco was used as the
medium of exchange, i.e. money, for almost two hundred years. After a
while it was deemed appropriate to issue receipts for tobacco that was
properly stored, these receipts thereafter being utilized as money, instead
of the tobacco itself. The tobacco remained in storage, while ownership of
the individual pack of tobacco changed when the receipt or storage slip
was used as a means of payment. Thus, the first step in the direction of
paper money was taken.

In contrast to precious metals, tobacco is characterized by the fact that
society can increase the amount of it relatively quickly. By allocating large
areas to tobacco-growing, the 'money supply' in society can be substantially increased. In order to prevent too rapid a growth of the money supply, the authorities in Virginia limited the area available for tobacco
cultivation.

Production of metals, on the other hand, is a more elaborate process.

This means that the money supply in the form of metals can increase only slowly over a period of time. As we shall discuss later in this chapter, too rapid a growth of the money supply will result in inflation.

12.2. Money as a Means of Payment

In a peasant community where each household generally produces what it consumes itself, there will be little need for trade. When production in a community becomes more specialized, there will be less overlapping between what the individual household produces and what it wishes to consume. In addition to the fact that increased specialization results in the growth of total production—as Adam Smith pointed out—specialization also leads to an increased exchange of commodities. Thus, a greater need will arise for an appropriate means of payment, i.e. money.

When 'something' is accepted as a general means of payment by everyone, transactions in the economy will be enormously simplified. To understand this point, let us imagine a market day in a small town. In the market-place thirty or forty stalls have been erected where individual producers have set out their goods for sale. However, the individual producer is also a consumer; she comes to the market with her own goods and wishes to return home with other goods. In other words, she is ready to barter to a considerable extent.

If our stallholder has apples to sell, and she intends to take home tomatoes, then she needs to find someone with the opposite preference: someone who has tomatoes and wants apples. The adjacent stall is full of tomatoes. However, the stallholder here is looking for cucumbers, so she is not interested in trading with the apple-stall. After a long search the two women find a third stallholder who has cucumbers and wants apples. Thus, a three-way transaction can be carried out. Stallholder 1 gives apples to stallholder 3, who in turn gives cucumbers to stallholder 2. And stallholder 2 gives tomatoes to stallholder 1. In this way they all end up with what they want.

In this story there was no generally accepted means of payment. If there had been, the process would have been much simpler. The apple-lady could have sold apples to anyone, as long as they had the generally accepted means of payment, i.e. money. Then she could have taken the money and looked around for the cheapest and best tomatoes, knowing full well that her money would be accepted there as a means of payment.

The general point of this simple example is that money makes it possible to divide transactions up. When apples, cucumbers, and tomatoes can be bought and sold for money, the individual stallholder can consider buying and selling as separate actions. She will not need to look for another stallholder with the opposite preference. The apple-lady's need to search for a tomato-lady looking for apples will cease. It is easy to see that the transaction will take place much more quickly when everyone in the market accepts, say, metal coins as money.

12.3. Money as a Store of Value and as a Unit of Account

A prerequisite for money functioning as a means of payment is the certainty that it will retain its value. In addition to the fact that money must function as a generally accepted means of payment, it must also function as a *store of value*. If our apple-lady does not spend all her income from the sale of apples on tomatoes (and other goods) one day, she can come back the next day, without apples but with money, and continue her shopping.

A third function money must have is what is called a *unit of account* or *numeraire*. By this is meant that all prices are measured in money. If the exchange value of apples in a market without money is measured sometimes against cucumbers, sometimes against tomatoes, and sometimes against potatoes, beans, onions, or whatever, an overview of what is expensive and what is cheap is easily lost. A smart stallholder will then be able to participate in a chain of exchanges where the lack of consistency between the various prices will be utilized to make a profit.

For example, if she is successful in bartering 2 kilos of apples for 1.2 kilos of tomatoes, 1.2 kilos of tomatoes for 1.5 kilos of cucumbers, and 1.5 kilos of cucumbers for 2.2 kilos of apples, she will make a profit of 0.2 kilos of apples. Such a chain of exchanges will be hard to carry out if all prices (for example of apples, tomatoes, and cucumbers) are in the same unit, e.g. copper coins. When the price of the various goods is measured against the same unit of account—money—the information concerning what the various goods cost will be much better. And it will be easier to compare prices and make sensible purchases.

To sum up, money has three functions: a generally accepted means of payment, a store of value, and a unit of account for what the various goods cost. If an economy is to function well, it is of prime importance that the monetary system works—in David Hume's words, that trade is 'well oiled'. Only when this is *not* the case do we understand the importance of a healthy monetary system.

12.4. Internal and External Convertibility

In many countries in Central and Eastern Europe today, the monetary system is out of order. Perhaps the greatest problems are in Russia and the other republics of the former Soviet Union. When prices determined by the central authority are generally kept lower than the prices required to equate the quantity demanded and the quantity supplied, i.e. when markets do not clear, the result is long queues outside the shops. When money alone is no longer sufficient to obtain goods and services demanded, the economy will regress to a pure barter economy. Rather than using roubles as a means of payment, people will learn that they can get further by using other goods as a starting-point for obtaining what they want. The function of money as a means of payment, a store of value, and a unit of account will be gradually eroded. In Supplement 6.3 John le Carré described how the protagonist in the thriller *Russia House* had to use tickets to the Philharmonic as a means of payment for imported soap which would subsequently be exchanged for the bolt of green check cloth of pure wool in a clothing shop.

When money no longer functions as intended, we say that it loses its *convertibility*; i.e. its 'exchangeability'. The oil that is supposed to make the wheels of trade run more smoothly becomes rancid. And economic activity falls below the level it is capable of attaining.

In Russia the rouble has largely lost its *internal convertibility*; it functions poorly as a domestic means of payment. Beside a 'rouble economy' we therefore find the gradual emergence of a 'dollar economy'. For the individual citizen, holding dollars is preferable to holding roubles. One can be more certain that dollars and other Western currencies will maintain their purchasing power over a period of time. In many contexts it is also easier to use dollars for domestic payments than roubles. When the prices of more and more goods are listed in dollars, this foreign currency will also gain increased importance as a yardstick for how expensive or cheap the various goods are.

By *external convertibility* we mean how easily a country's money can be exchanged for other countries' money (or currencies). In Poland, where over a twenty-year period the 'dollar economy' gradually invaded the 'zloty economy', the decisive step was taken in 1990. For imports and exports the zloty became freely convertible to the dollar, at the rate of 9,500 zloties per dollar. A widespread practice was thus fully legalized, and the black market trade in foreign currency dried up, quite simply because this trade was no longer illegal, and the official rate conformed with the black market rate.

One important purpose of pegging the zloty to the dollar was to re-establish confidence in the domestic currency. When zloties could be legally exchanged for dollars, the zloty became more suitable as a means of storing value and a means of payment. But if politicians are to be successful in the long term, inflation in Poland will have to be brought down to the same level that applies to the American dollar, i.e. to the US rate of inflation, currently 3–4 per cent per year. Otherwise a devaluation of the zloty will inevitably take place, and its purchasing power will be reduced. In May 1991 this was precisely what happened. Since price rises in Poland from 1 January 1990 had been far greater than those in America, the value of the zloty compared with the American dollar was reduced by almost 18 per cent. This meant that the price of a dollar rose from 9,500 to 11,200 zloties.

12.5. The Gold Standard

In the period 1880–1914 most countries had linked the value of their paper money to gold. This meant that, if so requested, each country's central bank was obliged to change notes into gold, at a fixed price, and vice versa. Rather than having a store of tobacco as security for the value of their banknotes, the central banks had to store gold. The quantity of notes they could issue was limited by the quantity of gold they had in reserve.

Since each country's currency had a fixed exchange rate in relation to gold, they also had a fixed price (or rate) in relation to one another. Thus, one American dollar could always be exchanged for 1.640 grammes of gold, while one British pound had a gold value of 7.988 grammes. Given these two exchange values in relation to gold, the rate of exchange between the two currencies was also set at $(7.988÷1.640) per pound; in other words, $4.87 was the price of £1 sterling.

If the level of costs and prices in the United States tended to rise in relation to the level in the United Kingdom, American goods would be more difficult to sell and the United States would have a deficit in its trade with the United Kingdom. Since payment for this deficit had to be made in the form of gold, the quantity of notes in the United States would have to be reduced. Lower circulation of money would in turn lead to lower prices. Thus, American goods would regain their competitiveness, and over a period of time the balance of trade would be restored.

The years 1880–1914 were characterized by considerable openness in the economic relations between countries, rapid economic growth, and the virtual absence of inflation. As a consequence of the First World War (1914–18), however, gold conversion was suspended. In the mid-1920s it was restored in many countries. However, as a result of the crisis in the international economy in the 1930s, gold conversion was again given up by all countries apart from the United States, where the central bank's right and duty to convert dollars to gold was retained until 1971. (After 1945 only other countries' central banks could demand gold for dollars.)

12.6. Printing Presses as Engines of Inflation

When the monetary system was disengaged from gold, the authorities experienced increased freedom to order their central banks to issue new notes. On the one hand, resources, in the form of extraction of gold that merely languished in a store as security for the notes, were saved. On the other hand, discipline in monetary policy was substantially reduced. When democratically elected governments in practice have a monopoly on the printing of new banknotes, it is tempting for them to resort to the printing press to cover increased public expenditure or to enable taxes to be cut. In the short term this will increase the government's popularity and therefore its chances of re-election.

In the short term, i.e. over a period of one or two years, inflation may be an undesirable consequence of the interplay between various economic and political forces even if the government has not printed extra money. However, lasting price rises over a longer period would not be possible without a substantial growth in the money supply.

In the United States in the 1960s, President Lyndon B. Johnson was unwilling to finance the unpopular war in Vietnam with increased taxes, as we mentioned in Chapter 9. When at the same time he increased public expenditure with a view to developing various domestic welfare schemes, he did so via a more rapid growth in the money supply than previously. To begin with things went fine. The 'oil'—money, that is—lubricated the 'wheels' nicely, and total production showed buoyant growth. However, when all resources were being utilized to the full, the relatively rapid growth in the money supply entailed higher and higher prices, i.e. inflation.

Throughout the latter half of the 1980s, a corresponding development took place in the Soviet Union as well as in other Central and Eastern European countries. Increased public expenditure, which was not possible to finance in the form of increased public revenue, was covered by the production of new banknotes. When there is faster growth in the supply of banknotes than in the supply of goods, inflation is on the cards. However, when prices are set centrally, the latent price rises will not be allowed to manifest themselves. The result is 'hidden' or 'suppressed' inflation. Actually, this is not so difficult to observe; the continually growing queues speak eloquently for themselves. So does the transition to pure barter and to trade based on other countries' money.

SUPPLEMENT 12.1. FIVE WAYS OF ELIMINATING SUPPRESSED INFLATION

Suppressed inflation occurs when the actual prices in the free market-place exceed the centrally stipulated ones. In such a situation people will have more cash and fewer goods than they wish. The transition from plan to market requires that this surplus stock of money must disappear in some way or other. Professor Peter Wiles of the London School of Economics and Political Science suggests the following five ways (which are not mutually exclusive):

1. Let the prices in the state-controlled shops rise so much that the queues disappear, as fewer people can afford the available goods. No wage compensation should be granted to allow for higher prices. To some degree this occurs in practice already—that is, when a good gets a 'new name' and a higher price, without any noticeable change in the product having taken place.

2. Institute monetary reform as in East Germany in 1990. For example, the quantity of roubles in the economy could be halved if two old roubles had to be exchanged for one new one. Another variant was carried through in the Soviet Union in January 1991. Through a decree issued by President Gorbachev, it was determined that each inhabitant be given an opportunity to exchange a limited number of 50- and 100-rouble notes for smaller denominations, and the remaining large notes were then declared worthless.

3. Increase production of consumer goods to the detriment of investment goods and military materials.

4. Sell off part of the state's gold reserves and spend the revenue on increased imports of consumer goods.

5. Let some of the goods in state shops be sold on the unofficial market (black market) at the high prices existing there. The roubles that accrue from this would then have to be removed from circulation—in other words, be incinerated.

Professor Wiles acknowledges that none of these methods would be easy to implement in practice (see Wiles 1988: 236–9).

For our part, we can submit a sixth method: let the state sell land, enterprises, houses, and flats to private persons. Such a way of reducing the circulation of money in the public sector would, however, require more than a decree by the country's president.

12.7. Relative Prices and the Absolute Price Level

We have previously discussed the difference between a change in relative prices and a change in the absolute (or general) level of prices (see Supplement 5.1). In a dynamic economy changes in relative prices will be natural. Changes in consumer preferences, i.e. in the composition of consumers' demands, will indicate the need for changes in relative prices. New inventions and new technology leading to more efficient production of various goods and services will pull in the same direction. In the West new technology has made refrigerators, computers, and watches cheaper and cheaper. However, increased wage levels have made labour-intensive goods and services, such as hand-made furniture, visits to restaurants, and repairs of various types, increasingly expensive.

If the money supply rises in accordance with total production, there will be nothing to prevent changes in relative prices from coinciding with stability in the absolute level of prices. It is only when the money supply over a lengthy period rises faster than production in society that the general level of prices will begin to rise.

12.8. Problems with Inflation

However, if money is the oil that lubricates the economy, what is wrong with too rapid a growth in the money supply? There are several problems connected with the inflation that follows from excessive monetary expansion. Perhaps the biggest problem is the distributional effect of price rises: those with monetary claims will lose; those who owe money will gain. When the general level of prices rises, money loses its purchasing power. Savings will buy fewer goods and services than individuals anticipated when they put the money aside. And those who owe money will get off more cheaply when debts are settled with money whose purchasing power has been reduced in the meantime. Those who have incomes stipulated in nominal sums and who are not in a position to change these—e.g. pensioners—will, during inflation, experience a reduction in purchasing power.

When a society gradually becomes accustomed to inflation, this will be reflected in interest rates, which are the price of borrowed money. If previously, in the absence of inflation, I was willing to lend out 100 roubles for one year against 2 per cent interest, then when inflation was running at 10 per cent I would demand a *nominal* rate of interest of 12 per cent. In both cases the purchasing power of the money lent will rise by 2 per cent. This means an interest rate of 2 per cent 'cleansed of inflation', or a *real* interest rate of 2 per cent.

SUPPLEMENT 12.2. INTEREST AND INFLATION

In this chapter and the next the concepts of interest and yield will be discussed. Let us say that the interest rate is 12% per year. By this is meant that financial capital (e.g. money in a bank account) grows for example from 100 to 112 roubles in the course of one year. But at the same time the general (or absolute) price level will probably also rise. Owing to inflation, therefore, it will be necessary in many contexts to distinguish between the *nominal* interest rate—i.e. the interest rate without taking the rise in prices into account—and the *real* or inflation-adjusted interest rate. In order to arrive at the real interest rate, one must quite simply subtract inflation from the nominal interest rate. If for example the nominal interest rate is 12% and the annual rise in prices is 10%, the real rate of interest will be (12 − 10)% = 2% per year.

A high nominal interest rate, which reflects an expected rise in price levels, will in turn cause problems. First, people will often have different ideas about what inflation in the future will be. If a borrower believes there will be a 5 per cent rise in prices in the next twelve months, whereas the lender expects 10 per cent, it will be difficult to agree on a suitable interest rate for the loan. Even if the borrower and the lender were to have the same expectations concerning future price rises, uncertainty surrounding inflation could have an inhibiting effect on both the desire to save and the will to invest. Moreover, experience has shown that it is hard to achieve a positive real rate of interest with a rise in prices of, let us say, more than 20 per cent per year. And a negative real interest rate will further inhibit saving in the economy.

Furthermore, a high nominal interest rate can cause problems with realizing long-term investments. If the revenues proceeding from a project come only some years after investments have been initially made, a high nominal rate of interest will require more and more new loans to pay the interest on the loan that was taken up initially.

With high inflation, it will be more difficult to distinguish between changes in relative prices and changes in the absolute price level. If I register that the price of shirts has risen by 15 per cent in the course of one year, it will not be easy to know whether shirts have become more expensive, relatively speaking, or not. If inflation this year was 20 per cent, shirts will have become cheaper compared with most other goods and services. But if inflation was 10 per cent, shirts will, relatively speaking, have become more expensive. The capacity of prices to give appropriate information will be reduced. This will lead to a poorer allocation of resources and reduced economic growth. So it is probably no coincidence that high inflation is associated with slow growth in many cases.

A final disadvantage of inflation which should be mentioned here is that a lasting inflation beyond a certain level (perhaps 6–8 per cent per year) has been shown from experience to undermine people's confidence in economic policy in general. When there is doubt about the future purchasing power of money, a general feeling of uncertainty and dissatisfaction may follow. This is scarcely a fruitful starting-point for a productive economy or a satisfied society.

12.9. Creating a Monetary System

The countries of Central and Eastern Europe are facing a formidable task when it comes to creating an orderly monetary system. In the Soviet Union in recent years government spending has exceeded government revenue by about 10 per cent of gross national product. Since this over-consumption by the government has been financed to a considerable degree by the printing of more money, while at the same time prices have been kept artificially low, a latent demand pressure of considerable dimensions has built up. The moment that prices are freed, they will experience a substantial one-time jump.

In this connection, it is necessary to consider two factors. First, a one-time jump in the level of prices does not have to be followed by a continuous rise in prices. If, thereafter, the authorities manage to keep the growth of the money supply under control, while at the same time a free-market economy is able to supply a steadily increasing quantity of goods and services, the one-time jump in prices may be followed by a stable (but of course higher) price level.

Second, the rise in the various prices will be far from uniform. With regard to goods where the excess demand arising from today's artificially low prices is particularly large, prices will rise more strongly than where excess demand is small or non-existent. And those goods that today are not saleable at the centrally stipulated prices will fall in price.

Thus, free pricing in an economy in transition from planning to free markets will have two elements in it: correct relative prices—what we were mainly concerned with in Part II of this book—and an increase in the absolute price level, as the rebuilding of a mismanaged monetary system will require.

SUPPLEMENT 12.3. THE UKRAINE INTRODUCES ITS OWN
MONETARY SYSTEM

On 1 November 1990 the Ukraine, the second largest republic in the
Soviet Union, introduced a monetary reform. Although this monetary
reform involved a wage reduction of some 30 per cent, people seemed
generally satisfied with it. Let us see what the reform consists of.

When workers receive their wages, in addition they are given coupons.

These coupons are not of the old type, which said '1 litre of milk' or '1 kilo of sugar'. For 100 roubles in wages, workers receive 70 coupons called 'karbovanets', which is the Ukrainian word for roubles. To buy something in the shops, they must pay with both roubles and a similar amount in coupons.

As people have gained confidence in this 'double' monetary system, their willingness to use money as a means of payment in buying and selling has increased. The result is more goods in the shops: meat, sausages, and milk are back on the shelves. The queues are still long, but at any rate there are goods to buy when you get to the counter.

One reason why this monetary reform has had the desired effect is that previously saved roubles are now losing much of their purchasing power. The surplus supply of roubles in the economy will be worth little when new coupons are also needed for payment. The very fact that the supply of coupons is so modest makes them valuable, so that people are willing to accept them as a means of payment.

If other republics were to introduce corresponding schemes, the Soviet monetary system based on roubles would be undermined. If each republic got its own coupon, one could quite simply get rid of roubles altogether. And the coupons would acquire the role of a monetary unit by themselves.

Source: After *The Economist* (15 December 1990): 64

13

A Closer Look at the Capital Market

Concern for profit is just what makes possible the more effective use
of resources.

Friedrich A. von Hayek

OVER time, economic growth is dependent on the accumulation of capital. This requires saving; i.e. the total income generated in the economy
must exceed the consumption of goods and services. Saving is a prerequisite for investments, such as the building of new production facilities and
the development of new and better machines. The concept of investment
is often extended to include the development of new *knowledge* in society.
Through education, knowledge and skills are acquired which will equip
the individual with a better production capacity. We then talk about
investment in human capital.

The capital market has an important role to play when it comes to
accumulation and allocation of the resources that society spends on real
investments, i.e. on extending production capacity. In this chapter we
shall give some examples of how people and enterprises make use of the
capital market. Towards the end of the chapter we shall discuss the capital market as a whole.

13.1. Joint Stock Companies

When a person saves, he sets aside some of his current income.
Consumption is renounced today in order to achieve greater consumption
later. If I have accumulated wealth amounting to 25,000 roubles, my
choice will be: either spend all the money on food, clothes, and holidays
in the coming period, or hold on to some of it as latent purchasing power.
Should I choose the latter, I will again face a choice: how to invest this
money in the meantime, i.e. how to *store my wealth*. Again we could imagine two alternatives: either put the money in the bank, or invest it in some
form or other of productive activity. If I put it in the bank I will earn
interest. If I invest it, for example in a shoe factory, the returns will be
more uncertain, but the possibility of making a far higher profit than the
interest the bank account will earn may make this alternative tempting.

Now, 25,000 roubles is not enough money to build a shoe factory. However, it is sufficient to buy shares or stocks in such an enterprise. If we imagine that there are twenty persons who are planning to establish a joint stock company with a view to manufacturing shoes, and each one of them has 25,000 roubles, the equity capital will be 500,000 roubles. If in addition we can borrow a corresponding amount from the bank, we may have enough money to implement the project.

Assuming that the other nineteen are prepared to invest immediately, what considerations must I take into account when they invite me and my 25,000 roubles to join them?

First, I must consider the plans that are presented. What sort of shoes are to be produced? What kinds of machinery and buildings are to be utilized? Who is to be in charge of the undertaking? What wages are the workers to get and how easy will it be to get hold of skilled personnel? Which suppliers of raw materials are available? How should the shoes be marketed? And what prices can we expect for the finished products?

The answers to all these questions can be spelled out in a *budget*. It will show expected costs and expected revenues, usually arranged on a monthly or quarterly basis for the coming couple of years. Before production starts up there will only be costs. After a while revenues will begin to come in. My interest in the project will naturally be dependent on the figures in the budget: the lower the costs and the higher the revenues, the more attractive the investment will appear.

13.2. Saving in Shares and Risk

With confidence, both in the nineteen other investors and in the budget presented, I accept the invitation, and place my hard-earned savings of 25,000 roubles at the disposal of the undertaking. Since we are all going in with the same amount, I will *own* 5 per cent of the new undertaking, or 5 per cent of the shares in it. Thus I will renounce the safe bank interest and choose the less secure investment, i.e. purchase of shares.

The months ahead will be exciting. Will the budgets hold, or are expenses going to be higher than planned? How are production and sales of shoes going? Are the style and quality up to consumers' requirements? Are we managing to sell the shoes at the expected prices?

If all goes well, over a period of time revenues will exceed expenses by a significant margin. I who own 5 per cent of the enterprise will then be able to enjoy a nice increase in the value of my shares. If on the other hand we have miscalculated, and the expected revenues do not materialize, I will risk losing the whole of my invested capital. However, I cannot lose more than that: share-owners can lose their share investments but not house and home. For this reason we say that a joint stock company is a company with *limited liability*, i.e. limited to the share capital invested. The owners of an enterprise can enjoy the revenues that are left after all expenses are paid. Wages to the employees, payment for raw materials and other input, and interest on loans from the bank are expenses that must first be met. Whether there will be any profit for the owners, only time will tell. We can therefore say that there is a greater risk attached to the funds the owners invest than there is for others who have economic demands on the enterprise.

Whether I should spend my 25,000 roubles on shares in the shoe factory will thus depend on my expectation of a greater profit than interest earned on a savings account. Otherwise I would not be willing to take the risk. On the other hand, I must be fully aware of the possibility that my whole investment could be lost. This will happen if the enterprise goes bankrupt. Even if the firm does not go bankrupt, however, if it does badly and has to dip into its share capital, after a couple of years I may end up with stock that is perhaps worth only 10,000 roubles. In such a case my starting capital will have shrunk to less than half.

When a joint stock company does well, the owners (or shareholders) can enjoy a handsome profit. This profit can be employed in two ways: payment of *dividends*, which we could say are a form of interest on the share capital, and a *rise in value* of the shares themselves.

As a precondition for there being any dividends to pay out, the enterprise must have made a profit after all other expenses have been paid. If this occurs, shareholders may nevertheless decide *not* to pay out dividends. Instead they may let the money remain in the enterprise and spend it on new investments which will result in increased production. In that case the shareholders believe that expansion of the enterprise's capacity will be a more lucrative way of utilizing their profits than alternative investments elsewhere.

13.3. The Shareholders' Role

Thus, share capital is one form of *equity capital*. In minor firms, such as the kiosk on the corner, equity capital may merely be the owner's own savings. None the less, as far as equity capital is concerned, what we have expounded above will apply: the owner of this capital will get what is left when all other expenses are paid.

There is a further specific property of equity capital: the right to decide how the enterprise shall be run. The kiosk-owner on the corner instructs his two sales assistants regarding selection of goods, opening hours, etc. He reaches agreement with them concerning wage conditions and duties.

In joint stock companies where there are several owners, the owners convene in a *general meeting*, normally once a year, where each owner votes, in relation to the number of shares he has. The general meeting elects the *board*, which in turn engages the *chief executive officer (CEO)* of the enterprise, or managing director.

There are no fixed rules concerning how many participants there should be on a board and how often it should meet. In our example with a small shoe factory we might perhaps indicate a board of four or five persons, and *board meetings* every second month. In the start-up phase, and when special problems or challenges arise, there will be a need for more frequent meetings.

The board, which should preferably be composed of people with differing qualifications, has the task of drawing up the long-term guidelines for the enterprise. Then it is the task of the CEO, who is also present at these meetings, to implement the various measures.

To sum up, we could say that the owners' role is to make their capital available, with the concomitant risk. As a reward for the risk, the owners receive any profits accruing. Furthermore, the owners have the right, directly or indirectly through the board and CEO, to decide how the enterprise is to be run. In negotiations with the employees, agreement is reached on wages and other working conditions; in talks with the bank, on loans and conditions for loans; and in contacts with suppliers and customers, on purchase and sale respectively of input factors and the finished products (shoes).

13.4. Return on Investment

Let us now assume that the shoe factory is doing well. In the first year it reaches break-even; i.e. total costs are equal to total revenues. The next year results in a nice profit of 150,000 roubles. Measured as a percentage of equity capital (which is 500,000 roubles), this amounts to 30 per cent for this year. You attend the general meeting where it is agreed that the profit should be ploughed in its entirety back into the enterprise in the form of new investments. The question arises which of two possible investment projects should be chosen. For the sake of comparison the CEO has prepared a short note on the two alternatives. This note is shown in Supplement 13.1.

SUPPLEMENT 13.1. A COMPARISON OF TWO PROJECTS

Our shoe factory has 150,000 roubles at its disposal and is considering what types of machines this money should be spent on. For that purpose I [the CEO] have prepared an overview of what additional revenues each project could be expected to result in.

For project A, an investment of 150,000 roubles this year would give an annual additional revenue of 120,000 roubles in each of the two coming years. For the two-year period as a whole this amounts to 240,000 roubles. The machine would then be worn out and have to be scrapped.

Project B is of longer duration: the additional revenue here would be 65,000 roubles in each of the next four years. The total additional revenue would therefore be 260,000 roubles.

At first sight project B would seem to be better than project A: 260,000 roubles in additional revenue seems more tempting than 240,000 roubles. However, here we must be careful. The additional revenue from project B would be spread further out in time than in the case of project A. We must find a method by which the projects can be compared and ranked better.

One way of doing this is to calculate the *return on investment (ROI)* for each of the projects. I have done this as follows:

Project A: $$150,000 = \frac{120,000}{(1+r)} + \frac{120,000}{(1+r)^2}$$

Project B: $$150,000 = \frac{65,000}{(1+r)} + \frac{65,000}{(1+r)^2} + \frac{65,000}{(1+r)^3} + \frac{65,000}{(1+r)^4}$$

These calculations show that the rate of return r for project A is 38 per cent and for project B, 26 per cent. Since project A, measured in terms of annual returns, gives more than B, I would recommend that A be chosen.

Project A is better than project B for the following reasons. The additional revenue from A for each of the first two years is 120,000 roubles, compared with 65,000 roubles for B. This means that if we choose project A, we will be able to realize more new investment projects in the enterprise earlier than if B were chosen. These new projects will in turn result in further additional revenue. If we take this into account, all in all, project A will generate more money than project B.

Another way of explaining these calculations takes as its starting-point the fact that one year hence 100 roubles will be worth less than 100 roubles today. If I can obtain 10 per cent interest, it is easy to see that 100 roubles today will be equivalent to 110 roubles in a year's time. This calculation can be turned around: given 10 per cent interest, in a year 110 roubles will be equivalent to 100 roubles today. The present value of 110 roubles one year down the road will be 100 roubles.

If the interest rate is set higher, e.g. at 20 per cent, the *present value* of 110 roubles in a year will be lower still (91.67 roubles). If we say that the profit requirement for investments in our shoe factory is set at 30 per cent, the present value (PV) of the two projects can be calculated as follows:

$$A: PV = -150,000 + \frac{120,000}{(1 + 0.3)} + \frac{120,000}{(1 + 0.3)^2} = 13,300$$

$$B: PV = -150,000 + \frac{65,000}{(1 + 0.3)} + \frac{65,000}{(1 + 0.3)^2} + \frac{65,000}{(1 + 0.3)^3} + \frac{65,000}{(1 + 0.3)^4}$$

$$= -9,200$$

Here we can see that the present value for project A is positive, whereas it is negative for B. Since only projects with a positive present value should be implemented, project A should be carried out but not project B.

The concept of return on investment (or internal rate of return), which has been summarily defined in Supplement 13.1, makes it possible to compare and rank different investment projects. Where projects are considered to have the same risk, one should quite simply select in accordance with the highest rate of return, as far as the money will go.

In the world of textbooks (and in Part II of this book), where free competition rules and all resources are in full use, without external effects, all prices will reflect *the best alternative use* of any resource. In such a world, ranking of investment projects in accordance with a falling rate of return will result in the best utilization of resources viewed from the enterprise's point of view as well as from society's point of view.

Irrespective of what one might think of capitalism versus economic planning, the return on investment is a useful concept. It shows how much more we will get back, calculated as an annual percentage return, by investing in the project in question today. In Western countries it is a requirement that state firms should also employ ROI calculations in the assessment of different investment projects.

As indicated above, in a planned economy where prices do not reflect the relative scarcity of resources (absence of scarcity pricing), it will be less meaningful to employ the rate of return as a criterion for which projects one should assign priority to. It is only when most prices in the economy are 'correct' that such calculations will gain validity. For this reason (among others), a *gradual* transition from planning to market will prove difficult—something we will return to in the final chapter of the book.

SUPPLEMENT 13.2. RANKING PROJECTS IN A PLANNED
ECONOMY

In a socialist planned economy, based on Marx's labour theory of value,
concepts such as profit and interest have little place. In terms of Marxist
theory, capital in itself is not productive. The use of concepts of interest
(such as the return on investment) to compare and rank projects with
differing time-spans is almost taboo.

Instead, a number of different indicators are used, such as the effect of
investments on consumption of energy and raw materials, without use of
prices or the effect of investments on total wage costs as a percentage of
total production. Another criterion for ranking of projects may be sums
invested per rouble in increased production. Often the problem will arise
that different indicators result in different rankings of two projects. Which
project is chosen in such a case will therefore depend on which indicator
is utilized. The whole thing will take on an arbitrary character, the more so
because prices and wages will seldom reflect the economic costs and
revenues to society.

Further, indicators used in a planned economy play down the significance of various types of service production such as transport, telecommunications, and various financial services.

13.5. Saving in Banks

Let us now return to our shoe factory and remind you that it received a
bank loan corresponding to the invested equity capital, i.e. 500,000 roubles. Thus, the bank has an *account receivable* or a claim on our enterprise.
The money we have borrowed from the bank is designated *loan capital* in
the enterprise's accounts and *receivable capital* in the bank's. The enterprise must pay interest on this money.

Should the shoe factory have problems surviving in competition with
other enterprises, the bank will not lose money before all of the share capital is used up. For this reason, loan capital carries a far smaller risk than
equity capital. As compensation for the greater risk, it is normal in a market economy for returns on shares, taken as an average for many companies over a period of time, to be greater than the interest on bank
deposits.

SUPPLEMENT 13.3. AVERAGE RETURNS ON SHARES V.
RETURNS ON BANK DEPOSITS

For the period 1967–87 the annual rate of inflation in Norway was 8% on average. In the same period the average return on (large) bank deposits was 11% per year. This means a *real interest* of (11 − 8)% = 3% per year. In the same period shares gave an annual return on average of 17% (composed of 5% in annual dividend and 12% in annual price rises for shares). The real return, i.e. the return measured in *purchasing power*, was therefore (17 − 8)% = 9% per year. Since, on average, shares gave 6 percentage points higher returns than bank deposits, we say that the market assigned investment in shares a *risk premium* averaging 6% in this period.

If we delve deeper into the figures, we will see that many companies went bankrupt in this period, and that far more occasionally experienced substantial falls in share prices. The person who owns shares in a period where the share price falls will lose money if he sells the shares at that time.

If we look at the period 1973–7, average returns over these four years were only 2% in the stock market, compared with 9% for bank deposits. In the period 1983–7 the picture was different: shares gave an average annual return of 28% compared with 15% for bank deposits.

The figures above show that people who invest their money in shares should be willing to take chances, and preferably should be able to afford losing it.

My neighbour, who has got a job in the shoe factory, is considering buying shares in it. He asks me if I am willing to sell mine. Since the enterprise has now existed a couple of years and is doing well, I think a suitable price would be 35,000 roubles for my 5 per cent stake. Recently another of the twenty initiators sold his stocks for this price. Both my neighbour and I agree that he can expect an annual return of at least 10 per cent on such an investment of his funds. The bank gives only half as much in interest on deposits.

For my part I am interested in freeing the money, partly for increased consumption, and partly to participate in a new project: establishment of a shirt factory.

After having given my offer some consideration, my neighbour decides that he does not want to buy the shares after all. It is not that I am demanding an unreasonable price. No, it is the assessment of the risk that is worrying him. For if the enterprise—against expectations—should start to do badly, he would risk losing not only his job but also the savings he spent on the shares.

Basically, my neighbour is probably just as interested in buying shares in the shirt factory that is being started. He thinks that expected returns and risk are approximately the same here as for the shoe factory. The reason he nevertheless would rather buy shirt shares than shoe shares is that his total economic situation would be less risky in this way. Why?

The reason is that he will be receiving his income from two sources: *labour income* from the shoe factory, and *capital income* (in the form of dividends and share price gains) from the shirt factory. If the shoe factory should go to the wall, it will not take the shirt factory with it. The value of the shares in the shirt factory will therefore not be affected by what happens at his daily workplace.

If my neighbour does not want to take any risk with his savings, he will end up depositing them in the bank. Then he will receive 5 per cent annual interest, and that will be that. Then it will be the bank's job to consider how the money should be utilized.

SUPPLEMENT 13.4. WHAT IS A STOCK EXCHANGE?

As examples in the text we have two firms. One produces shirts, the other shoes. In a market economy of the Western type there are thousands or tens of thousands of joint stock companies. Most of these are small and have only a few owners, for example a family. However, there are also large undertakings with several thousand shareholders. Naturally, all of these shareholders cannot know one another personally, much less all those who are interested in becoming shareholders in their company. In order to enable share-owners who are interested in selling to find buyers (and vice versa), stock exchanges have therefore been established. A *stock exchange* is quite simply a market-place for the purchase and sale of shares.

Assuming that many transactions in the stocks of the various companies take place each day—which happens only for large undertakings—it will be possible each day to register the prices of the various shares, i.e. the *share prices*. By weighing the share prices together daily in an index, it will be possible each day to register how the agents in the stock market as a whole assess the future.

Such indexes are published daily in financial newspapers such as the *Financial Times* in London. The most well-known is the Dow Jones Index of the New York Stock Exchange. If an optimistic mood predominates as far as all the registered companies' future profits are concerned, the share price index will go up (called 'hausse' in French terminology). If instead a pessimistic mood reigns, and many people believe there will be hard times, the index will fall ('baisse').

13.6. The Capital Market Viewed as a Whole

After this examination of the way in which my neighbour and I evaluate how we should utilize our saved capital, let us now take a closer look at the capital market from the point of view of the economy as a whole. Initially we stated that the capital market has important tasks to perform concerning the accumulation and allocation of capital in society. We have also made a distinction between equity capital and loan capital. We shall now introduce the distinction between real capital and financial (or receivable) capital.

Real capital in a society is the stock of physical production equipment that exists at a given point in time. Machinery and buildings, as well as railways and ships, are typical examples. Such objects perform productive services over a period of time. They are gradually worn down and must be replaced by new ones, if production is not to fall. If investments in society over a period of time are greater than those that are necessary to replace old equipment, we say that net real investments in the economy are taking place. Then the way will be paved for economic growth.

The *financial capital* in a society is the sum of all the receivables or monetary claims that exist in the economy. If as a starting-point we take a country without trading links with other countries, the sum of liabilities and the sum of receivables will be equal in size. For society viewed as a whole, net receivables or net financial capital will therefore be equal to zero. Why? Because if I lend you 10,000 roubles, thus being registered with this sum in receivables, you will be registered with the same amount in liabilities. For society as a whole, receivables minus liabilities, i.e. net receivables, will be equal to zero.

When any liability is equivalent to a corresponding receivable, we will also see that gross loan capital in the economy will necessarily be equal to gross receivable capital. The 500,000 roubles that were designated in the shoe factory's accounts as loan capital would thus be designated as receivable capital in the bank's.

Viewing the economy as a whole, it might seem as if what happens to receivable capital is of far less importance than what happens to real capital. So it would be, too, if those who made investments in real capital financed these investments mainly with their own savings. That is not the case. When a new commercial undertaking is initiated and when established undertakings expand, usually the need will arise to borrow money in addition to the equity capital the owners are able to raise. (The shoe factory discussed earlier in this chapter required 500,000 roubles in share capital and just as much as a loan in the bank.)

When the decision to save is separated from the decision to invest in production capital, the workings of the capital market will be of considerable importance. Two questions will arise: how to ensure that total savings in society are sufficient, and how to see to it that these savings are allocated efficiently among alternative investment projects.

The decision regarding how much of his current income the individual will choose to set aside as savings depends on two factors: how much he can afford to save, i.e., how big his income is; and what the possibilities for returns are. In an economy that manages a successful transition from planning to market, profitable investment projects will mushroom. And enterprising persons will initiate new undertakings where profitability appears to be best.

In the first part of this chapter we offered you the opportunity of entering a joint stock company for production of shoes. If you have previously been able to get only 5 per cent interest in the bank, and can now envisage a return of 30 per cent, it is entirely possible that your desire to save will increase. Over a period of time, when the market supplies more goods and services and the long queues outside the shops are a thing of the past, you will have a further incentive to save: namely, the certainty that funds saved today can be converted into goods and services later.

13.7. The Role of Banks

With many profitable projects on the drawing-board, the shortage of capital will be reflected in higher interest rates. In Supplement 13.1 we showed the calculation of the internal rate of return for two projects in our shoe factory. Other enterprises will have corresponding plans, and many of them will go to the bank to ask for loans to implement them. The bank will then have the important task of selecting which projects it should lend money to. Since there is not enough money for all of them, the bank will increase the rate of interest on its loans. This means that the less profitable projects—those that have a lower rate of return—will drop out of the loan queue.

In its evaluation of those remaining, the bank must make a critical examination of plans and budgets before loans are granted. In many cases the bank will extend loans, knowing full well that there is a possibility of the project failing. On loans which the bank regards as being rather more risky than usual, it is reasonable that it will stipulate a higher rate of interest, i.e. demand a risk premium.

When the bank increases the general interest rate on loans, it will also follow this with higher interest on deposits. This will tend to stimulate the public's interest in saving in banks. And with increased deposits, the bank will have more funds for new loans. In this way the economy will get into a *virtuous circle*; many profitable projects will be developed, interest rates will increase, saving will increase, and the best projects can be realized. However, a caveat is in order. With too high an interest rate on loans, the banks will end up holding too many risky loans. Thus, a somewhat lower rate of interest than what clears the market, i.e. a certain degree of *credit rationing*, can take place in a competitive credit market.

The banks' role in a market economy is twofold. First, banks guarantee the deposits of those who wish to save. A bank deposit is really nothing more than my loan to the bank, corresponding to the bank's debt to me. Bank deposits must therefore be characterized as *loan capital* from the bank's point of view.

The banks' other task is to grant loans to those who wish to invest. The banks' own profits lie in the difference between interest on loans and interest on deposits. This difference is called the *interest margin*.

In addition, banks can convert short-term deposits to long-term loans. This means that, while I am free at *any* time to withdraw the money I deposit in the bank, the person who borrows it may have a clear agreement to pay back the loan over, let us say, a period of five years. If the bank is reasonably sure that most depositors are stable savers, there will be little risk involved in letting short-term deposits finance long-term loans. In times of crisis, however, when confidence in a bank is shaken, such a risk may arise: fearing that the bank will go bankrupt, everyone will run to the cash desk to withdraw their money.

Early in the 1930s there were such runs on many banks in the United States. The result was that almost a quarter of the American banks went bankrupt. Those depositors who did not get their money out lost their savings. This served to prolong the Great Depression, making the 1930s an even harsher period.

In retrospect, most economists agree that the US Federal Reserve, i.e. the central bank that is responsible for the issue of new notes, and functions as a bank for the other banks and for the government, did a poor job in this period. If the 'Fed' had been more willing to lend money to banks that got into trouble, the crisis would have been less widespread.

Having learned their lesson, central bankers reacted wisely when, in October 1987, the value of shares on the New York Stock Exchange on Wall Street fell by $1,000,000,000,000 (i.e. $1 trillion) in the course of one week. The Fed declared immediately that the banks would be provided with sufficient liquidity in the wake of the stock price plunge. So there was no run on banks, no financial panic, and no economic crash.

Having learned from experience, the central banks of the West have much greater responsibility for the overall health of the banking system. This means that a country's central bank—in some cases in collaboration with other public authorities—checks that the legal requirements concerning equity capital in the individual bank have been satisfied. Further, the individual banks' accounts are carefully scrutinized by the supervisory authorities. They make sure that the accounts give a true picture of the bank's economic situation.

Proper supervision of the banks' operations gives depositors greater security concerning their deposits. In the United States bank deposits up to $100,000 are insured by the authorities. However, deposits exceeding this amount recently vanished at the stroke of a pen, in connection with the bankruptcy of a small bank in New York. If a depositor with a large sum in her account feels uncertain about how dependable the bank really is, she should withdraw her money and deposit it in another bank.

Sound banking requires deposit insurance to be accompanied by proper bank supervision by public authorities. Without proper supervision of their loans and lending policies, banks tend to be attracted to risky projects which promise high returns if they succeed, for if they fail the depositors' money is insured by the authorities. Deposit insurance without proper supervision is, in effect, a licence to gamble with public funds and an invitation to catastrophe. The gigantic savings and loans débâcle in the United States is a case in point: the ongoing bail-out of bankrupt financial institutions at taxpayers' expense is estimated to cost the American public in the end as much as $500,000,000,000, equivalent to $8,000 for each family of four in the country, and perhaps more.

In a well-developed capital market banks and savings and loans associations are not alone: there are a good number of other financial institutions. Pension funds, insurance companies, and unit trusts are among the most important ones. They supply various financial services. With a broad range of such services from competing institutions, customers— enterprises as well as individuals—can avail themselves of a variety of financial services.

13.8. Summary

The capital market consists of two parts: the market for equity capital (such as shares), and the market for loan capital (such as bank deposits). Since decisions concerning savings are often separated from decisions concerning investments, the capital market has two important tasks: accumulating savings, and allocating them to investors. The rate of interest— which is the price of borrowing or the reward for saving—will affect both total savings in the economy and what projects will be realized.

Furthermore, the capital market has an important task as redistributor of risk. This applies at two levels. First, the person who is willing to take a risk can utilize his own money as equity capital, either by starting his own undertaking or by buying shares in someone else's. The more careful person will deposit his money in the bank. Thus, he leaves it up to the bank to decide who in turn can borrow this money.

Second, the capital market makes it possible to store wealth in an undertaking that is separate from one's daily place of work. We remember the neighbour who, having evaluated the overall economic risk, did not want to buy shares in the shoe factory where he himself worked.

This second point is worth keeping in mind when state enterprises are to be privatized in Central and Eastern Europe. In this connection it is common to argue that the employees should be given a significant portion of the shares. This is intended to have a motivating effect on their work effort; as owners the workers will enjoy part of future profits. In many cases this will certainly be right. In the West, too, there is an increasing interest in workers' shares and worker-managed enterprises.

However, two objections are worth putting forward. In a large enterprise the individual's work effort will mean little to the total profit of the company. The motivation of owning a few stocks will therefore be modest; although the symbolic effect—and that can be important enough—will perhaps be greater than the pure economic incentive. The other objection was that of the neighbour in this chapter: his overall risk will be less when the returns on his savings have nothing to do with his workplace.

14

The Labour Market in a Competitive Economy

In the labour market, as well as in other markets, industrialists and firms strive to achieve the best possible results for themselves. This means that owners of firms and stockholders want to keep wages low. Employees of course have the opposite wish: to get as high a wage as possible for their work effort.

In the long run adjustments in this market take place in the same way as in other markets: both the individual enterprise and the individual employee must on the whole accept the existing prices (here, wages) and adapt to this situation as best they can. The reason why the labour market deserves a special chapter in this book is that wages in modern market economies are largely determined in *negotiations* between employees' and employers' organizations. An exception to this general rule is provided by the United States, where to a much greater extent wages are determined by the interplay of supply and demand in labour markets without so much interference from nationwide organizations of labour and business.

Karl Marx envisaged the powerful employer standing before a 'reserve army' of workers. Then he could pick and choose among the best and reject the others. Competition between workers for jobs would force wages down and set the workers against one another. However, today's labour market in most industrialized Western countries looks quite different.

14.1. Labour Market Organizations

Even before the turn of the century, workers had become organized in *trade unions*. To this day it is an important task for organized workers to persuade their non-union comrades to join the union. The more members there are, the stronger will the union be in its negotiations with the employer regarding wage rates and other working conditions.

The development of trade unions in this century has been closely connected with political developments, but this need not concern us much here. Our main objective in this chapter is to discuss the economic aspects of trade unionism and labour relations. At the risk of oversimplification, we shall assume that unions have two major aims: safe jobs and high wages. As we shall see later, these two objectives may come into conflict with one another. But unions are not alone in labour markets. Some time after the trade union movement had established itself, employers began to organize themselves in *employers' associations*.

14.2. Strikes and Lockouts

In some Western European countries there has traditionally been close co-operation between trade unions and the country's leading labour party (or social democratic party). At times this has caused political problems. If considerations concerning inflation and competitiveness in relation to other countries result in a labour party government wishing to cut back public expenditure, the trade union movement may often wish to resist this. This is because in the short term such a policy of cutbacks will result in increasing unemployment and lower wage increases. If the trade union movement is strong enough to force the hand of the government, the problem may be postponed, but with the grave danger that it will be even greater later on. Less current unemployment will have been achieved in 'exchange' for greater unemployment in the longer term.

In order to persuade employers to give in to their demands, trade unions can threaten to *strike*. In many cases the threat will be enough. However, now and then a strike will break out; i.e., workers will refuse to show up for work.

There are examples of short one-day strikes supported by only a few persons as well as of strikes affecting a large number of people lasting for months. In normal times, when there are no conflicts, members make regular payments into a *strike fund*. If there is a strike the workers receive pay from the savings in this fund. How long a strike can last will thus be dependent upon the size of the strike fund.

In order to increase the strength of trade unions during a labour dispute, different unions will often collaborate. One union may commit itself to contributing funds from its own strike fund to the members of another union, if this should prove necessary. In this way the endurance of the strikers will be strengthened. However, unions will not make their strike funds available in this way without demanding in return the right to exert influence over decisions concerning the strike.

An alternative way of giving support is the *sympathy strike*. For example, a strike at shoe factories may be supported by transport workers, who will refuse to transport the stock of finished shoes from the factories as long as the strike lasts.

In the face of such threats employers do not have to remain passive. Their weapon is the *lockout*, whereby the workers are barred from their workplaces and lose their wages. In the short term the employer will also be hurt by this. However, if he knows that the workers have low strike funds and that their demands are unpopular among other unions, he may nevertheless decide to give measure for measure.

Employers too can support one another in conflicts. In the same way as trade unions, employers' associations build up funds during normal times. Enterprises that experience strikes or lockouts can get financial support from these funds. In this way the losses for owners of enterprises that are drawn into labour disputes are limited.

An employers' association can also impose *fines* on its member-enterprises. An enterprise that pays its employees more than what has been agreed on in central negotiations can be fined by its own association.

In strikes or lockouts the wheels of industry remain motionless, and the country suffers an economic loss. For society as a whole, therefore, strikes and lockouts are expensive weapons to use in a dispute. On the other hand, these instruments of power give the parties an incentive to reach an agreement. If significant societal interests are threatened by a widespread labour dispute, legislation in some countries gives the authorities the right to intervene, for example with *compulsory arbitration*. In such cases a publicly appointed commission, in which representatives of both sides of the labour market are represented, will stipulate binding agreements between employers and employees which shall apply during the coming period between central wage negotiations.

14.3. Organized Functionaries

In Marx's classical ideology, only workers organized themselves in unions. In a modern European type of industrialized country, *functionaries* ('white-collar' workers) are also organized.

Both workers and functionaries can be organized in several different local (or national) unions, even in the same enterprise. Often the criterion for such organization is the members' *occupation*. For example, foundry workers may be organized in one union, locksmiths in another, accountants in a third, transport workers in a fourth, etc. Such organization by occupation is found among other places in Denmark, Iceland, and the United Kingdom. In one and the same enterprise—if it is large enough—workers may be represented by ten or more different labour unions.

Another criterion for membership in a labour union may be the *branch* or *industry* one works in. For example, everyone working in the metal industry may belong to one and the same union, everyone in the mining industry to another, etc. This is the typical pattern in Germany, France,

Italy, Norway, and Sweden, for example. This results in far fewer trade unions in the same enterprise—perhaps only one for workers and one for functionaries. A third variant, which we find among other places in Switzerland and Japan, is based on each *enterprise* having its own trade union. When employees are organized in this manner, wage levels can vary among enterprises in the same branch or industry.

No matter how the labour market is organized, competition for labour in periods characterized by low unemployment will force those enterprises that pay their personnel poorly to raise wages. Some will not manage this and will have to close down. However, with a great shortage of manpower the possibilities of better paid jobs in other enterprises will be good. In periods where there is a fear of increasing unemployment, variations in wages from enterprise to enterprise will make it easier to reduce unemployment. This is due to the fact that the less profitable enterprises will retain their manpower even if they pay their personnel less than the more profitable ones.

The fact that functionaries have organized themselves is due to the market economy's gradual transition from mainly producing goods to more production of services. A large and increasing proportion of functionaries are, therefore, to be found in the public sector; in other words, local and central government are their employers.

SUPPLEMENT 14.1. WAGES AND UNEMPLOYMENT

One of the most serious flaws of modern market economies is their intermittent failure to guarantee full employment of labour. Most recently, the 1980s and the early 1990s have been a period of substantial and persistent unemployment in many countries of Western Europe. The Great Depression of the 1930s is another case in point.

Earlier in this book, especially in Chapters 3, 5, 9, and 10, we argued that unemployment cannot always be eliminated quickly or painlessly. In particular, we showed in Chapter 3 that unemployment results from wages being kept above the level that would equilibrate labour demand and labour supply (see Fig. 3.5). We asked: 'Why do wages not fall so that all who are willing and able to work can get jobs?' We mentioned the balance of power among employed and unemployed workers as a possible explanation: those whose jobs are safe have no interest in seeing wages fall, and they will resist attempts by unemployed workers to offer their labour at lower wages rather than remain without jobs.

Another possible explanation of persistent unemployment, mentioned in Chapter 3, deserves some further comments here. Consider a factory which pays its workers good wages in return for competent labour services rendered. Suppose also that there is some unemployed labour available; outside the factory gate there is a queue of people willing to work at lower wages. Under these circumstances, the owners of the factory may be tempted to try to cut costs and raise profits by lowering wages. What would happen if they did?

They might lose some workers to other firms, and they would then have to replace them with new workers who would have to be trained at substantial cost. Moreover, the workers who would stay on despite lower wages would be less satisfied and probably less productive than before. So, all things considered, cutting wages may not be a cost-saving measure. On the contrary, it may reduce profits by reducing labour productivity. Therefore, despite unemployment, the firm may not have any incentive to lower wages. Hence, joblessness will persist.

14.4. The Public Sector

We can observe a further two departures of modern trade unionism from Marx's classical ideology: (1) functionaries are a substantial group within the trade union movement; and (2) the public sector is an important employer.

Both of these factors will affect the workings of the labour market, viewed as a whole. In the individual enterprise workers and functionaries have the same counterpart, i.e. the employer. However, at the same time workers and functionaries will compete with one another concerning division of total wage payments. Both unions will therefore be in a quandary. On the one hand they will have an incentive to stand together against their employer. On the other hand they will find themselves competing with one another; it will be important to ensure that benefits obtained for one's own members are at least as great as those that are achieved by other unions.

Experience has shown that competition between different labour unions is assigned considerable significance during wage negotiations. In many cases unions are more concerned about their members' relative wage development than about ensuring that workers and functionaries viewed as a whole receive the greatest possible proportion of the total value added in the enterprise.

In a society where the public sector is a significant employer, strikes will be a two-edged sword. The public sector never goes bankrupt. On the contrary, when for example state employees strike, when teachers, train-drivers, and health personnel stay at home, this saves public money. Rather than affecting the employer, strikes in the public sector hit a 'third party': people who have children in school, who are dependent on trains, or who need nursing.

Unavoidably, politicians are drawn into complicated labour disputes in the public sector. Why? Politicians have two roles. They are the elected representatives of the people, and in that role they have a need for a reasonable degree of popularity to ensure re-election. On the other hand, politicians are also employers in the public sector. These two roles will readily come into conflict. As an inflexible employer, the politician may provoke strikes. Then a 'third party'—the man in the street who also has a vote at the next election—will also suffer (schools closed, trains at a standstill, nurses at home). On the other hand, if the politician in the role of employer gives in too easily to excessive wage demands, the result will be inflation. Over a period of time too rapid a rise in prices can have unfortunate consequences for the country's economy viewed as a whole.

In the public sector, therefore, neither strikes nor lockouts are particularly attractive instruments for employees and employers. The battle concerning wages and working conditions will therefore be fought primarily in the media; the party that gains most sympathy for its own viewpoints on radio, TV, and in newspapers will often do best in negotiations.

In the appeal to public opinion lies much of the challenge of presenting one's own demands as coinciding with a 'just' solution, without specifying what criteria justice should be measured by. Against this background it is easy to understand that wage disputes—especially within the public sector—are often characterized by intense lobbying among journalists, demonstrations aimed at TV, frequent readers' articles in newspapers, etc. Such input is important in the struggle to win the support of public opinion.

In Western market economies, with a public sector that employs one-fourth to one-third of the total labour force, there are no simple solutions to the dilemma surrounding the dual role of politicians. Now and then various proposals are submitted aimed at introducing various types of 'watertight compartment' between the role of employer and the role of democratically elected representative. Ultimately, however, it will be the same politician who decides how tax revenues will be utilized, no matter whether these revenues are allocated to higher wages for public-sector employees or to other things.

Thus, we can conclude that reality demonstrates a third important departure from Marx's classical ideology: the influencing of public opinion and the role of the mass media in labour disputes in an open democratic society.

14.5. The Rights of Employees

A fourth and final difference that should be mentioned here concerns various types of rights for employees, often introduced through legislation. For example, employees in a number of countries have the right to appoint one or more board members in their firm. Further, employees have a right to information on changes at their workplace and an opportunity to present their viewpoints in connection with reorganization of the firm. There are various formal procedures which protect employees in connection with layoffs, harassment, or dismissal because of trade union work. Via legislation, regulations governing holidays, confinement leave, the organization of the working environment (among other things for reasons of safety), etc., have also been introduced.

Viewed as a whole, a Western type of labour market is quite complicated. There are a number of trade unions which, depending on the situation, sometimes co-operate, sometimes compete. The public sector is taking an increasing proportion of the total employed labour force. Under the influence of mass media, politicians in such a situation are frequently weak in their role as employers. As a result, unions have often tended to have the upper hand in negotiations with employers' associations in the past. The willingness of many governments to accommodate excessive wage increases—by, for example, devaluing the currency or increasing public spending to prevent an increase in unemployment following wage negotiations—has contributed to this outcome.

14.6. Wage Settlements in Practice

In connection with wage settlements, representatives of the trade unions and the employers meet for negotiations on wages and working conditions for the coming period—usually one to three years ahead.

Many factors will influence the wage agreement that is finally signed. If inflation in the current period has been greater than expected, the employees, through their unions, will normally demand compensation for this in the form of higher wage increments. If the increase in prices has resulted in increased profits in most enterprises, chances are good that the unions will succeed in their demands. The enterprises will then be able to afford to raise wages. However, if the price increases are due for example to energy and imported raw materials having become more expensive, the enterprises will be less willing to give compensation for inflation.

The situation in the labour market is another factor that will influence wage negotiations. In times of heavy and increasing unemployment, employers will be unwilling to pay higher wages. In such a situation, where the supply of labour exceeds the demand, the market will indicate that wages ought to decrease. The unions, too, may moderate their demands if there is high unemployment. In such a situation higher wages could reduce the firms' demand for labour, thus leading to a further increase in the number of people out of work. On the other hand, an increase in total wage payments would lead to increased purchasing power among the workers; this in turn could stimulate demand in the economy, and thus production and employment too.

Unions will often produce this last argument. However, if firms react by reducing the work-force when wages rise, rather than increasing production, the result will be more people out of work. In that case those who keep their jobs will enjoy higher pay at the expense of those who lose theirs.

The conflict between safe jobs for everyone and high wages for those who keep their jobs is obvious here. And the solidarity within the trade union movement will be put under strong pressure.

In the Scandinavian countries there is usually a national collective agreement which is settled centrally between the major organizations in the labour market (trade unions and employers' organizations). This agreement is usually followed by local negotiations at the individual enterprise. Those who work in well-run enterprises can therefore obtain wage increases in excess of the centrally stipulated ones. Such increments can also be negotiated in the course of the period covered by the agreement, and be made dependent on development of productivity in the enterprise. The collective agreement may also contain rules providing a framework for such local bonuses.

In addition to wages, it is possible in collective agreement negotiations to take up questions such as weekly working hours and holidays. In an economy experiencing substantial economic growth, it will be reasonable to expect that employees will take out some of this growth in the form of more leisure time.

14.7. Wage Growth and Competitiveness

An important factor in the limitation of wage growth is competition from abroad. Firms that grant their employees greater wage increases than those of workers in competing enterprises abroad may run into problems after a while. If growth in production per worker (growth in productivity) is the same at home and abroad, greater wage increases at home will force prices up. In the short term a reduction of competitiveness may perhaps be limited by reducing the profits that accrue to the capital-owners. In the long term, more rapid growth in wages at home will lead to higher prices for our goods compared with the goods supplied by our foreign competitors. This will result in increased problems in exporting and selling our goods abroad and in a better competitive situation for foreign imports to our domestic market. All in all, our production will slacken, layoffs will be unavoidable, and our total profits will be reduced.

To summarize, both employers and wage-earners in firms exposed to foreign competition will know that, if they permit too great a wage increase, their undertaking will risk pricing its products out of the market. In the final analysis this will affect employees in the form of unemployment, and may result in increased joblessness and even in bankruptcy.

A corresponding mechanism for ensuring moderation in wage levels does not exist for employees in the public sector. State-employed teachers, train-drivers, nursing staff, etc., do not work in competition with foreign enterprises.

If the wage structure in an open economy is to function, therefore, all parties in the labour market must accept that those unions whose members work in enterprises exposed to competition from abroad are wage-setters for the whole economy. Then the norm for wage developments in the public sector will be set by enterprises that exist in competition with foreign producers. If public-sector employees are given higher wage increases than employees in enterprises exposed to competition, experience has shown that a combination of envy and competition between different trade unions will result in wage increases in the public sector carrying over to the exposed sector. The result will be reduced competitiveness and increased unemployment.

For this reason, politicians must attempt to strike a balance between two objectives. On the one hand, they must seek to restrain wage growth in the public sector so that an excessive wage increase here does not spread to the private sector and threaten competitiveness and jobs. On the other hand, they need to guarantee local and central government employees conditions that are good enough to attract well-qualified people to jobs in the public sector, in areas such as education, health care service, and public administration where talented personnel are needed.

SUPPLEMENT 14.2. A LESSON FROM JAPAN?

The experience of many Western European countries suggests that the labour movement has succeeded in exerting an important influence on wages over the years, as intended. This is not surprising in view of the fact that about one-half of the labour force in Britain, France, Germany, and Italy is unionized, compared with one-sixth in the United States, for example. Therefore, wages have generally tended to be more dependent on the relative bargaining strength of unions and employers' associations, and less sensitive to demand and supply in individual labour markets, in Western Europe than in the United States. This means that employment and wages have also been affected by strikes during the centralized bargains between the labour market organizations in Western Europe. Indeed, joblessness can sometimes be traced to a wage bargain that exceeded the ability of firms to pay, resulting in layoffs.

Why would unions push for a wage increase that results in unemployment? One reason is that unions safeguard the interests of a majority of their members, sometimes at the expense of the unemployed minority. Moreover, unemployed workers are less likely to belong to unions and to be able to influence their policies than those who have jobs. Thus, workers whose jobs are secure can benefit from pricing others out of work, inadvertently or deliberately. This problem arises because unions, whether they are organized by occupation or by industry, generally represent different types of workers employed by different firms with different abilities to pay.

Fixed wage payments tend to contribute to unemployment, in both Western Europe and North America. Suppose an export company's sales decline because of a recession abroad. To maintain profitability and sales, the company must cut costs and prices. Labour costs are frequently an overwhelming part of total costs. When wages are unrelated to profitability, i.e. when employers and workers have agreed on fixed wages, the quantity exported is reduced to maintain prices and profitability. Production declines and workers are laid off.

Labour markets in Japan are largely free from these flaws. There, as a rule, each firm negotiates directly with its own workers. The workers have an incentive to keep their wage demands within their firm's ability to pay because otherwise they may jeopardize their own jobs, not somebody else's. In addition, base salaries in Japan are generally supplemented by bonuses related to the profits of the firm. This system of profit-sharing enables many Japanese firms to adjust to a decrease in sales by lowering bonus payments rather than reducing employment. That way the burden of adjustment is shared by all: everyone loses a little; nobody loses everything.

This seems to be an important part of the reason why the Japanese economy has been characterized by considerably less inflation and less unemployment than most other industrial economies since the end of the Second World War.

Source: After Weizman (1984)

15

International Trade

The sole effect of high duties on the importation . . . of corn . . . is to divert a portion of capital to an employment, which it would not naturally seek. It causes a pernicious distribution of the general funds of the society—it bribes a manufacturer to commence or continue in a comparatively less profitable employment.

David Ricardo

SINCE the end of the Second World War international trade has blossomed as never before. For the world as a whole, international exchange of goods in 1960 constituted 8 per cent of total value added; thirty years later the figure had doubled to 16 per cent. The Western market economies have taken the lead in this development. Through the dismantling of customs barriers and other trade impediments, international exchange of goods and services has been encouraged. Increased specialization has further contributed to stimulating trade between countries.

International trade in countries with planned economies has not grown anywhere nearly as rapidly. Before we look more closely at trade in a planned economy, we shall give a short introduction to theories of international trade.

15.1. Adam Smith on Absolute Advantage

Why trade? Why not aim for national self-sufficiency? More than two hundred years ago, Adam Smith provided a definitive answer to these questions. He concluded that international exchange of goods was a source of wealth for all countries. Trade provides the possibility of increased specialization and better utilization of each country's economic resources.

A simple example will illustrate Adam Smith's point. Let us assume that England uses 1 hour's labour to produce 1 metre of cloth, and 2 hours to produce 1 litre of wine. If we disregard other production factors, this will lead to 2 metres of cloth being equal in price to 1 litre of wine. In that case, 2 hours' work will be exchanged for 2 hours' work.

Imagine that the situation is the opposite in Portugal. Here 1 hour's work is needed per litre of wine and 2 hours' work per metre of cloth. In that case 2 metres of cloth will have a price equal to 4 litres of wine in that country. (Then 4 hours' work will be exchanged for 4 hours' work.) All of this is summarized in Table 15.1. It shows that an hour's work produces 1 metre of cloth and half a litre of wine in England and half a metre of cloth and 1 litre of wine in Portugal.

TABLE 15.1. Hours of work in England and Portugal: Adam Smith's story

	England	Portugal
1 metre of cloth	1 hour	2 hours
1 litre of wine	2 hours	1 hour

Since the cloth manufacturer in England gets 4 litres of wine by selling 2 metres of cloth in Portugal compared with only 1 litre of wine at home, he will have every reason to sell his cloth in Portugal. Correspondingly, the wine producer in Portugal will have an incentive to sell his wine in England—he will also get a price four times as high for his product here. (At home he will get a ½ metre of cloth per litre of wine, in England 2 metres of cloth.)

In this situation, profit-maximizing producers will ensure that an international exchange of goods is arranged. Perhaps the price will finally be 1 litre of wine per metre of cloth. In that case both countries will have reason to be satisfied. There would be no talk of one country exploiting the other one. The trade is voluntary, and both countries will profit from it. The English will increase production of cloth and the Portuguese will increase production of wine. With a price relationship of 1:1 they will both supply goods corresponding to 1 hour's work, and receive goods equivalent to 2 hours' work.

In this example the English were more productive in cloth production and the Portuguese in wine production. In other words, they each had an *absolute advantage* in their area. However, all that is necessary to establish trade is that, before trade becomes a reality, the *exchange relationship* between the two goods is different in each of the countries: 2 litres of wine per metre of cloth in Portugal, compared with a ½ litre of wine per metre of cloth in England.

15.2. David Ricardo on Comparative Advantage

In the beginning of the nineteenth century David Ricardo took the theory of trade a long step further. He let England be less efficient than Portugal

TABLE 15.2. Hours of work in England and Portugal: David Ricardo's story

	England	Portugal
1 metre of cloth	3 hours	2 hours
1 litre of wine	6 hours	1 hour

in production of *both* goods: more labour is required to produce both cloth and wine in England than in Portugal. Table 15.2 shows the point of departure for Ricardo's assessments.

With 6 hours of work available in each of the countries, Portugal will manage to produce either 6 litres of wine or 3 metres of cloth. This means that the exchange relationship between the two goods is 2 litres of wine per metre of cloth. In England, 6 hours' work can give either 1 litre of wine or 2 metres of cloth. The exchange relationship here will thus be ½ litre of wine per metre of cloth. Since the exchange relationship between the two goods is different in the two countries, trade that is profitable for both parties will be established. Through specialization and trade, both countries will be able to make a profit.

The two tables were constructed in such a way that in both of them wine was four times as expensive (measured in cloth) in England as in Portugal. In Ricardo's story, where England is less productive than Portugal, England will nevertheless have a *comparative advantage* in production of cloth: although in absolute terms the English are less efficient in production of both goods, *relatively speaking* they are more efficient at producing cloth than wine.

Ricardo's major contribution lay in his proof that it was the domestic price relationships before trade that was the decisive element, not efficiency in the production of each good. A country that is less efficient than another country over the whole range of products may nevertheless be able to make good profits through specialization and international exchange of goods.

This insight may seem counter-intuitive; surely, one must be particularly efficient to export successfully? No, says Ricardo, and no, says reality: all that is necessary for mutually profitable trade is that the country be more efficient in relative terms in one branch of production than in another.

When trade between countries has actually been established, conditions may be right for further advantages; Englishmen specialize in production of cloth, and over a period of time they may develop new technology which results in increased productivity. The same applies to production of wine in Portugal.

Total production of the two products in the 'world' (here: England and Portugal) will increase for two reasons: the static one, i.e. exploitation of comparative advantages, and the dynamic one, i.e. specialization and increased productivity over a period of time.

However, we ought to note that making way for international trade is not without problems. Wine producers in England (and cloth manufacturers in Portugal) will have every reason to object to free exchange of goods. For them this trade will result in reduced production and less income. If the economy as a whole is to gain from the opportunities that the free exchange of goods over national borders offers, *reorganization* of production will be required. In the short term this may be a painful process, but in the long term it will be a profitable one.

In the theory of international trade that has emerged in the past 150 years, factors other than relative differences in productivity have been adduced to explain comparative advantages (and thus different exchange relationships between the two goods in the two countries prior to trade). Without going into depth about such theories here—after all, this is only an introductory book on the transition from a planned to a market economy—we can point out that differences in the relative supply of factors of production can in many cases explain international exchange of goods and services.

For example, it is no coincidence that the Soviet Union, with its rich supplies of oil and gas, became an exporter of energy. Approximately 40 per cent of the Soviet Union's earnings in hard currency were due to such exports. A developing country, which is equipped with a plentitude of labour and little capital, will find it profitable to export products that require a relatively large amount of manpower in production, whereas industrialized countries rich in capital will specialize in the production and export of capital-intensive goods.

With such division of labour the developing countries will get an *indirectly* increased supply of capital; indirectly they will receive capital which is 'baked into' the goods they import. And correspondingly, the relative shortage of labour in the industrialized countries will be indirectly reduced through imports of labour-intensive products from the developing countries. In this way trade in goods will function as a form of substitute for international mobility of labour and capital.

Finally, trade in the same type of commodities (e.g. cars—see Supplement 7.1) constitutes a steadily increasing share of the international exchange of goods. Different preferences of people, the advantages of large-scale production, product differentiation, and the importance of branded products can explain such intra-industry trade.

15.3. Whence Protectionism?

In the sections above we have seen that free trade is a source of wealth; firms increase their productivity through specialization, making it possible for wage-earners and their families to buy less expensive goods and services with higher incomes than under a protectionist trade regime. Still, despite a gradual liberalization of international trade since the end of the Second World War, protectionist tendencies remain firmly ingrained in Western market economies: various existing government-imposed trade barriers like tariffs and import quotas remain. How can this be explained?

The explanation has to do with the uneven distribution of the gains from and cost of government protection. The owners of a firm threatened by import competition obviously gain from protection of the firm. The same is true of the workers in the firm, at least in the short run. The two groups of beneficiaries are concentrated, visible, and easy to identify. So also is their potential human and economic suffering if their firm cannot compete and if they do not quickly find new jobs.

The firm's owners and workers will have a strong incentive to expend both effort and resources in order to ensure protection for themselves. In a society with active media and ongoing competition among political parties, the firm's precarious situation will be a public issue. Pressure will be brought to bear on public authorities to 'do something' about the situation.

Political decision-makers can indeed do a few things—but protection against healthy competition is not the right answer.

Many decision-makers actually seem to be unaware that protection imposes a burden on consumers and other firms. This is partly because it can be difficult for them to obtain accurate estimates of the cost of protection. This is unfortunate. Such cost estimates are more abstract arguments in political life than the potential suffering of a vocal, visible, and relatively small group of people.

But are there none to defend free trade and point out the overall gains to society from free trade? Who protects the public from protectionists? The answer is: a precious few, for two main reasons.

First, consumers who stand to lose most from protection are typically an ill-organized, amorphous mass of people. Each of them has many other things to worry about than just the price of a particular commodity or group of commodities. So the consumer's incentive to organize a pro-free-import lobby is much weaker than that of producers (i.e. owners plus workers). It would also be expensive to do so.

Second, the cost of protection is generally rather low to each consumer, especially when considering one commodity at a time. But when consumers' costs are added up, the implicit transfer from consumers to workers of the protected industry frequently turns out to be very high. For various countries and commodities these costs have been estimated. It has been found that the cost of saving a job in the textile, clothing, steel, and motor car industries often amounts to some $US50,000–100,000 per worker and year. One reason why the cost is often very high is that the price not only rises on the imported quantity, but spreads to domestically produced substitutes. In fact, the latter price rises are the source of the benefits to some domestic producers at the consumers' expense. But these costs are typically hidden. A tariff is usually quoted in percentage terms, along with a quota in number of units allowed to be imported per year. It can be quite difficult to calculate the cost to consumers, especially when the protection takes the form of a quota.

To take the argument one step farther, suppose we want to prevent the workers from becoming unemployed. Then, is protection not a good thing after all from a purely distributional point of view? No, because there are more efficient, less wasteful ways of supporting the individuals whose jobs are at stake. Protection is primarily a support to production, not to individuals. When we are worried about individual well-being, we should support the individual as directly as possible to get to the root of the problem. In this case it means income support, retraining facilities, support to the individual if he wants to move to another region which lacks workers, etc.

When there is a general tendency towards increased unemployment in the country, selective protection against foreign competition is not the appropriate measure; protection just moves jobs from one sector of the economy to another. Since consumers would have to pay more for, say, protected textiles, they would have less to spend on other goods. Workers employed in the production of these other goods would, as a result, be fewer than otherwise. Employment in the clothing industry would also suffer since its input—textiles—would become expensive as a result of the protection.

Instead, in a situation with an economy-wide rise in unemployment only general macroeconomic measures can be effective. Through such measures an overall slack in employment can sometimes be eliminated, but not always, as we saw in Chapters 5 and 10. An example of such a measure is a devaluation of the currency, which makes *all* imports rise in price domestically and thus helps import-competing industries at home. A devaluation also makes *all* exports less expensive abroad and more profitable at home, thus benefiting domestic export industries. When appropriate, a devaluation is less discriminating and more effective than protection, other things being equal.

Further, in a situation characterized by full employment, it will be much easier for anyone threatened by unemployment (e.g. owing to fierce foreign competition) to find a new job. Maintaining full employment in so far as possible, and thus creating alternative job opportunities through macroeconomic and individual-specific microeconomic policies (e.g. labour market policies), may be the most effective anti-protectionist policy when all is said and done.

A small country must also reckon with the reactions from the rest of the world. The possibility of retaliation must be a matter of concern for small economies like the Central and Eastern European ones. A small country that increases its tariffs or introduces quotas will face protests from other countries. In the end, the country's exporters can be hurt by reduced access to foreign markets. Sometimes exporters form a pro-free-trade pressure group. The objective is to liberalize foreign trade, for instance through multilateral negotiations within the world's free-trade organization, the General Agreement on Tariffs and Trade (GATT).

In practice, new protection will be almost impossible for the Central and Eastern European countries to introduce. However, they can take different views regarding the pace at which existing trade barriers should be dismantled. Today Poland is possibly the most free-trade-oriented country of Europe, having completely overrun its domestic (potential) protectionist pressure groups before they realized quite the extent of protection they had been enjoying. This, of course, could lead to increased short-run unemployment in Poland. Hungary, on the other hand, takes a more gradual approach, which carries with it the risk of strong industry pressure groups being formed for the purpose of fighting trade liberalization later.

15.4. Planning-Directed Trade

How, then, have countries with planned economies exploited the advantages inherent in international exchange of goods and services? Let us first take a brief look at their trade with market-based economies, and then say a few words about trade between planned economies.

Planned economies have utilized the opportunities for profit through trade with market-based countries very poorly. Let us point to three self-imposed problems in the Eastern Bloc countries' trade with the West.

The first problem concerns incentives for the individual enterprise. The desire that the individual enterprise manager may have to engage in exports can hardly be motivated by the need to live a quiet life. In connection with exports to the West, stricter demands on quality than for domestic supplies are generally made. And foreign currency revenues accruing from sales abroad must be immediately handed over to the authorities. Further, the invoice price entered into the enterprise's accounts will seldom be set higher in connection with exports than for domestic sales. The demand for higher quality will therefore not be compensated for through pricing. Finally, if the enterprise encounters problems with sales abroad, the superior planning authorities will express their displeasure.

Under these circumstances, the desire to engage in exports to the West—to the extent that this takes place—must be due to other advantages to the management than profitability alone. For example, the opportunity to travel abroad in order to negotiate agreements may have been an important spur, particularly when private persons were ordinarily prohibited from leaving the country.

Another self-imposed problem is the incapacity to exploit the comparative advantages in the economy. In the absence of scarcity pricing, it will not be easy to know what the true costs of producing the various goods and services are. If we return to the example in Table 15.2 and let Russia take Portugal's place, with a simple device we can illustrate this point. The device consists of letting Russia utilize taxes and subsidies in a way that discriminates between wine and cloth producers. For, if the sale of cloth in Russia is heavily subsidized while the sale of wine is correspondingly burdened by taxation, then the price relationship between the two products prior to trade may be so powerfully distorted that Russia stands forth as a country with a comparative advantage in production of cloth. In that case the country will end up by *importing* the product for which it has a comparative advantage (i.e. wine) instead of exporting it. The consequence—in this simple example—will be that Russia will supply goods that demand *more* resources (measured in hours of work) than those it receives. And then the economy as a whole will suffer a loss instead of making a profit on international trade and commerce.

SUPPLEMENT 15.1. COMPARATIVE ADVANTAGE IN PRACTICE

Based on domestic prices in Russia, we can roughly say that 5 tons of copper in Russia can be exchanged for 1 car.

If we look at the international prices of copper and of a corresponding car, the relationship will be: 2 tons of copper = 1 car.

For Russia this means that, if the country reduced car production by 1 unit and increased copper production by 5 tons, these 5 tons could be used in exchange for 2½ cars. If Russia thus exploited its comparative advantage in the production of copper, an increase in copper production of 5 tons would give a net profit of (2½ − 1) cars = 1½ cars.

This calculation, however, is based on a precondition that may not be valid, i.e. that the resources used in production of 1 car could be used *alternatively* in the production of 5 tons of copper. In an economy based on scarcity pricing, this would be the case. When Russia goes from a planned to a market economy and scarcity pricing becomes a reality, Russians will gain reliable knowledge concerning which goods and services the country has comparative advantages in.

Source: After *The Economist* (20 October 1990): 10

The fact that, with erroneous domestic prices, international trade can easily lead to losses in the national economy rather than profits is more than a theoretical oddity. Calculations from Romania show that the country has sold refined oil products at a price of $US25 *less* per ton than the country paid for the crude oil that is used in this production (Winiecki 1988: 154). In the words of *The Economist* (5 January 1991: 55), 'Valuable metal, plastic, cardboard, rubber, energy go in at one end; Trabant cars worth less than the sum of these parts emerge at the other.' For (the former) East Germany, this means that sales of metal, plastic, etc., at international prices would have given higher revenues than sales of Trabants. Society would have been better served if the manpower used in production of these cars had played bridge instead. Production in this case was a process that *reduced* the value of the raw materials that were utilized.

15.5. Trade with Western Countries

The third self-imposed obstacle to utilizing the opportunities for profit inherent in specialization and international exchange of goods lies in the Eastern Bloc countries' express objective, as a group entity, of being as independent as possible of the West in all areas—that is to say, as self-sufficient as possible. One consequence is concentration on a relatively broad range of products, with little possibility of becoming equally efficient in all areas. The possibility of specialization, with an attendant increase in productivity, will thus be limited.

Another consequence is that production for export to the West was previously determined only on the basis of the need for imports. Necessary inputs which the Eastern Bloc countries did not have supplies of themselves, some luxury goods for the privileged and food for the masses, as well as imports of technologically advanced products, all had to be paid for in the form of export revenues in hard currency. With the demand for imports so determined, the task at hand was to ensure correspondingly large exports. Ricardo's thinking, that trade with other countries provides an opportunity for exploitation of comparative advantages and thus can contribute to better utilization of the country's resources, did not have a prominent place in this theory. In accordance with Marxist theory, work in service industries, such as trade, was looked upon as unproductive activity.

This led to a special form of export behaviour. Let us assume that the overall plan dictated that the demand for export revenues in hard currency for an Eastern Bloc country was $US20 billion in the coming year. These export revenues were broken down into individual sectors of the economy and then into individual enterprises. Let an enterprise that was instructed to earn $US10 million based on exports of agricultural machinery experience an unexpected increase in the demand for its products. The plan specified that 1,000 machines were to be exported at an expected price of $US10,000 per unit. Even if the demand abroad exceeded the expectations reflected in the plan, the enterprise had no interest in selling more than planned. For if 1,200 machines were sold at $US11,000 per unit, the enterprise would risk having to earn more revenue the following year. And then the market situation might be far more difficult—one can never tell what competing capitalist producers in other countries might get up to.

Given that the plan focuses largely on dollar earnings by the enterprise, the result may rather be that the quantity exported will *fall*. If a price of $US11,000 is obtainable, one might be content to sell only 910 machines abroad and still earn the $US10,000 imposed by the plan. It might be appropriate to hold the resources that were intended to be used on the last 90 machines in reserve for various purposes; compare the discussion in Chapter 6 on the need of enterprises to have reserve stocks of most things.

To summarize, self-imposed obstacles, such as a lack of incentive in the individual manager, the absence of scarcity pricing, and the desire for self-sufficiency, will make it difficult for centrally planned economies to extract the profits that international trade with market-based economies offers. Isolation from the Western market economies will also involve heavy restrictions on the possibilities for transfer of technology.

15.6. Bilateral Balance in Trade

How, then, has trade between the Eastern Bloc countries functioned? Have they had a more appropriate system? Unfortunately, trade organized through the CMEA (Council for Mutual Economic Assistance)[1] has contributed only modestly to enabling each country to exploit its comparative advantages. An important reason for this is that trade within the CMEA has been based mainly on *bilateral balance*. This means that each *pair* of countries seeks to balance exports against imports. This was one of the reasons why the *CMEA* was dissolved at the beginning of 1991.

The fact that the extent of international trade will be reduced when bilateral balance is demanded can easily be illustrated. Let us assume that Poland has good possibilities of a substantial surplus in exports to Romania, while trade with Hungary looks like resulting in a corresponding deficit. If trade with both Hungary and Romania, viewed in isolation,

---

[1] The CMEA, also called COMECON, was established in 1949 and consisted of the Soviet Union, Hungary, Czechoslovakia, Bulgaria, Poland, Romania, Albania, and East Germany. In 1962 Albania left and Mongolia came in. Later Cuba and North Vietnam joined.

must balance, Poland must reduce its exports to Romania and its imports from Hungary.

In Poland's trade with the West, on the contrary, there is no such limitation. If Poland has a deficit in trade with France, this will not be a problem if the surplus in trade with Britain is of the same dimensions. The reason is that French francs and British pounds are both *convertible currencies*. This means that these two currencies can be freely exchanged for one another. Since the French accept payment in British pounds, the Poles can settle a deficit in trade with France through a surplus in trade with Britain.

In 1963 the 'convertible rouble' was introduced. The purpose was to replace bilateral balance with increasing multilateral balance between the *CMEA* countries. In that case a trade surplus that Poland might have in relation to Romania could be used as payment for the deficit with Hungary. However, since the prices in the trade between *CMEA* countries often deviated from prices in Western markets, interest in using this system decreased. For let us assume that the goods Poland exported to Romania had a higher value for Poland if they were sold to Britain. The Poles would then be more interested in reducing trade with the East and instead increase trade with the West. The Romanians on the contrary would be most interested in buying as much as possible from Poland, as the goods they got from there would be cheaper than those from Western markets. Thus, the system of convertible roubles never came to function in accordance with its purpose.

The absence of scarcity pricing, both within each planned economy and between them, thus constituted an effective obstacle to specialization and exploitation of comparative advantages.

For goods for which it is hard to find Western prices—often because the quality in countries with planned economies lies far below that which is demanded in market-based economies—the prices in CMEA trade were agreed through negotiations. When in addition it was necessary for trade to balance between each pair of countries, the whole thing degenerated into pure barter. Poland was willing to sell a given quantity of a given product to Romania only if, in exchange, it received something of corresponding, or preferably higher, value to the Poles.

Owing to the absence of convertible currencies between the Eastern Bloc countries, money was not allowed to do its job as settlement for trade between these countries. We are back to the apple-stall-holder in Chapter 12 who, in the search for tomatoes, had to look around for a stallholder with tomatoes who was hunting for apples.

Since the CMEA system has now fallen apart, the way is paved for the old planned economy countries to increasingly utilize hard, convertible currencies in their mutual trade. In this way they will release themselves from the demand for bilateral balance, and will be able to benefit more from international division of labour.

SUPPLEMENT 15.2. MULTIPLE EXCHANGE RATES

In 1937 the cost of an American dollar was 53 kopeks. In 1950 Stalin stipulated the price of a dollar at 4 roubles, despite the fact that his economic advisers had recommended 14 roubles. Stalin chose quite simply to erase the figure 1, because he wanted his own currency to look strong on paper. 'This made no sense at all,' say Shmelev and Popov (1989: 229), 'since the rouble exchange rate thus set was not used in any accounting.'

In trade with Western countries, the exchange rate (the price of dollars measured in roubles) was determined on the basis of the kind of goods in question.

For example, when exporting lumber, the foreign trade associations convert the hard currency received into roubles according to one exchange rate while, when they import equipment, they receive hard currency for roubles at an entirely different rate. (Shmelev and Popov 1989: 229)

In reality, there were more than 10,000 different rouble prices for the American dollar in Russia. The capacity of prices to function as a signal for allocation of scarce resources completely evaporates under such a policy.

PART IV

Privatization, Accounting, and Challenges for the Future

THE introduction of a market economy necessitates private ownership of the means of production. The transfer of ownership from the public sector to the private sector raises many difficult issues. It is important that the new (private) owners utilize the means of production efficiently. Moreover, privatization, not only of enterprises but also of dwellings and arable land, must be conducted in a way deemed fair by most people. Considerations of efficiency and of fairness may easily clash and, if so, an appropriate balance must be struck between the two.

In order to make rational economic decisions, the managers of enterprises need information on costs and revenues. An appropriate accounting system is thus required. In the day-to-day operations of a firm, budgets are also needed. By comparing the actual figures on income and expenditures with those of the budget, management will get an early warning when things go astray.

The transition from plan to market entails huge challenges. Price reform and free competition presuppose private ownership and a dismantling of barriers to trade. Competition from abroad curtails the power of domestic monopolies. Prices fall. Customers can choose among alternative suppliers. This ensures a more efficient utilization of resources.

In the transitional phase, however, many companies will be driven out of business. Thus, an increase in unemployment for a while seems difficult to avoid. For the population to accept the inevitable economic hardships which the transition to a well-functioning market economy is bound to entail, it is of vital importance that the governing authorities enjoy political legitimacy in the eyes of the general public. Thus, the establishment of political pluralism and democracy and the transition from plan to market may be viewed as two sides of the same coin.

16

Privatization in Central and Eastern Europe

> The Theory of Communism may be summed up in one sentence:
> abolish all private property.
>
> Karl Marx and Friedrich Engels

PERHAPS the most important reform necessitated by the transition from plan to market in Central and Eastern Europe will be privatization of the property that was nationalized under communism. Reforms in this area will demand a parallel solution to a number of political, economic, and legal problems. In this book the emphasis is on the economic considerations.

In the main, four different types of property will be privatized: small businesses, homes, farms, and large enterprises.

16.1. Three Aspects of Privatization

A politico-economic problem, in this case the question of privatization, can often be considered in terms of three aspects: efficiency, distribution, and stability.

Efficiency How should ownership be organized so that the existing production apparatus and real estate are utilized most efficiently in the short term (often called 'static' efficiency)? A similar question concerns efficiency in connection with investments in new production capacity and in real estate, i.e. efficiency in the long term (often called 'dynamic' efficiency).

Distribution How is the distribution of income and wealth influenced by alternative methods of privatization? Here, too, there is a shorter perspective (distribution of income) and a longer perspective (distribution of accumulated income, i.e. of wealth).

Stability How are macroeconomic variables such as inflation and unemployment influenced in the short run by alternative means of privatization? And how will structural readjustment over time be affected by the process of privatization?

TABLE 16.1. Three aspects of privatization: efficiency, distribution, and stability

	Short term	Long term
Efficiency	1 Static efficiency	2 Dynamic efficiency
Distribution	3 Distribution of income	4 Distribution of wealth
Stability	5 Adjustment to cyclical fluctuations	6 Structural adjustment

Table 16.1 provides a way of organizing the arguments, but no guarantee that all the areas can be filled with clear answers. Often the answers in only one or two areas will be crucial for the decision as to how privatization should be implemented in a particular instance.

16.2. Small Businesses

The reason why small businesses have been placed in a separate category is that a small business can be sold relatively easily to only one new owner. The definition of a 'small business' in several Central and Eastern European countries is precisely that one person can own it. This may apply to shops, hairdressers, small hotels, restaurants, lorries, taxis, small factories, repair shops, etc. In Poland and Hungary small businesses are defined as undertakings where the role of owner under communism was generally exercised by local authorities. Thus, the question of privatization of small businesses in these two countries is to all intents and purposes a municipal affair.

Small businesses can be auctioned off. This method has been employed in Czechoslovakia. The person who thinks he is best qualified to run the business at a profit will usually make the highest bid at the auction. A public auction with many participants will increase the probability of getting 'the right person in the right place', i.e. the most suitable person as owner.

However, there are a number of problems even in connection with this relatively simple form of privatization. We shall look briefly at three: consequences of price reform, problems of financing, and foreign buyers.

Serious potential buyers will need to make an estimate of expected costs and revenues before submitting an offer for the small business. In order to carry out this calculation they will have to guess the future prices of the business's raw materials (inputs) and finished goods (outputs). This can be hard enough in a market economy; in Central and Eastern Europe there arises the additional problem of evaluating the consequences for future profitability of current and coming price reforms. Consider, as an example, an enterprise that today sells subsidized outputs (e.g. foodstuffs) and utilizes subsidized inputs (e.g. electricity); when the subsidies are cut back substantially or eliminated completely, the market prices that will then emerge may differ considerably from today's prices. The structure of future prices will not necessarily be known to anyone.

As far as internationally traded goods are concerned, to begin with it will be possible to use international market prices and add-on transport costs, customs charges, etc. This will give a good starting-point for calculation of future prices. However, of the inputs and outputs that are important to a small business, probably only a minor part will be the object of international trade. Then one may consider the market prices of so-called 'non-tradables' (i.e. goods that are not, or are only to a small extent, traded internationally) in other (Western) countries, such as energy and cement: these market prices could be employed in one's own computations instead of today's subsidized prices, and an estimate of future profitability made.

When subsidies are removed, prices will rise. Fewer people will be able to afford the goods, so demand will fall and queues will begin to disappear. However, the producers, who have now been freed from the overall plan, and who see that the prices of their goods are rising substantially, may wish to increase production. Use of overtime may become necessary, and investments in new machinery may appear profitable. For some goods—where queues were previously long and subsidies relatively low— consumers will show great willingness to pay more. This means that the market can tolerate both increased production and higher prices. For other goods (with short queues and large subsidies) demand may become smaller when prices increase sharply; however, increased production of these goods will gradually force prices down. In Poland just such increases and decreases in prices of a number of goods took place in the first months of 1990, following the deregulation of prices on 1 January.

With considerable uncertainty surrounding the future equilibrium prices in Central and Eastern Europe, it will be almost impossible not to cost enterprises on erroneous price estimates. This means that more careful but perhaps more qualified persons may at first refrain from buying and running businesses.

In addition, there is a dynamic aspect of the question of efficiency in an economy characterized by subsidies and regulated foreign trade. From the point of view of the economy as a whole, it will be important that new investments are based on correct long-term prices. Otherwise society will risk ending up with unprofitable investments and, moreover, with influential groups that resist necessary price reforms, in an attempt to protect their own recently implemented investments.

In other words, both static and dynamic efficiency speak for an early price reform and for freely established market prices for non-tradables.

Another question concerns the financing of the purchase of a business. In addition to his own funds (equity capital), a potential buyer of a business will often need to borrow money (loan capital). In a market economy he will then approach a bank with an application for a loan. The bank will consider the business concept, in this case the purchase of the enterprise. If the bank is convinced that the business concept is sufficiently sound that the loan can be repaid with interest, it will make funds available. In today's Central and Eastern Europe, however, there is no market-oriented banking system. Qualified bankers and financial analysts, who could assess the various investments and thus ensure efficient capital utilization, are also very much at a premium.

A third question relating to the sales of state-owned small enterprises concerns the extent to which foreigners should be allowed to buy them. From the point of view of efficiency, foreign owners would be a good thing. They would contribute experience and expertise concerning technology and management. In addition, new foreign owners would bring capital into the country. This imported capital could consist partly of Western machinery and other equipment, and partly of the Western currency the new owners would have to pay to take over the enterprise.

16.3. Agriculture and Housing

Restructuring of agriculture entails two factors: division of agricultural undertakings into smaller units, and privatization. If a large farm is divided up into smaller plots, there may be less economies of scale. This by itself could reduce efficiency. On the other hand, experience from the division of large land properties into smaller ones, owned by those who till the land, indicates that the result will be a rapid and substantial growth in production. Since both state-owned and collective farms in Central and Eastern Europe are enormous in size, this would indicate that the division and privatization of the land would be the best way to increase and improve food supplies. In connection with privatization, it is reasonable, also from considerations of efficiency, to give a certain degree of preferential treatment to those who are cultivating the land at present; they will know the specific preconditions for production (soil conditions, climate, etc.) and thus in general should be able to utilize the land most efficiently.

Concerning housing, not least blocks of flats, experience has shown that people who own their own flats are more interested in spending time and money on maintenance and renewal than people who merely pay rent. (This phenomenon was discussed briefly in Chapter 3.) This has proved to be the case in countries in Central and Eastern Europe where houses and flats are now owned by those who live in them. In connection with privatization of the housing market, therefore, the way should be paved for home ownership. However, there will also be a need for a rental market. In order to meet this need the state might find it appropriate to sell blocks of flats to private investors. They in turn would rent out the flats, expecting to obtain good profits from the funds invested. From a dynamic efficiency viewpoint, it would then be important to reduce or completely remove the regulation of rents. If rents are determined by the authorities, the owner of a block of flats will have no incentive for renovation or maintenance—since such expenses cannot be compensated for in the form of higher rent. The incentive to build new dwellings will also vanish if the public authorities are required to stipulate the price of rental.

16.4. Large Enterprises

The privatization of large enterprises raises completely new issues. From an efficiency viewpoint stiff competition between many enterprises is desirable. Existing monopolies must be broken up, and other obstacles to competition removed. Where possible, large enterprises should be divided into smaller units and then privatized. Further, it is important to make conditions as favourable as possible for establishing new businesses. As we have previously observed, free trade with foreign countries is important, both in order to guarantee correct prices at home, and so that customers will have alternative suppliers to choose from. Customers' choice is an expression of real competition between producers.

If in addition foreign enterprises are allowed to establish their own production plants in the country, conditions for greater competition and more efficient utilization of resources will be further improved.

In purely general terms, it will be important to have business legislation that encourages competition, including regulations that safeguard open and non-discriminatory public-sector purchasing, a law that ensures free access to establish new undertakings, and a ban on price and bidding cartels.

One consequence of the aim of obtaining maximum profits is that wastage of resources will be minimal. Another is that satisfaction of customers' demands will be central to the enterprise's operations. While strong competition will force the management of a large enterprise into behaving in such a desirable way, professional owners (i.e. the shareholders) will reinforce such behaviour.

Share ownership can in principle be exercised in two different ways. The active way will involve shareholders' monitoring the company closely. If the enterprise is badly run, the owners will force changes upon it, for example by replacing the management and electing a new board. This active way of exercising ownership is often called the 'voice method'; the owners let their voices be heard in protest against poor operation.

According to the passive way of exercising ownership, those shareholders who are dissatisfied with the way the company is being run will sell their shares, for example on a stock exchange. This method is called the 'exit method'; the owners' dissatisfaction is shown by the fact that they choose to leave the sinking ship. If many shareholders use the exit method, the price of the shares will fall. This will cause dissatisfaction among the remaining (and new) shareholders. They will have an incentive to shake things up in the management of the company, thus causing changes to take place in the operation of the company (the voice method). Furthermore, a lower price for the company's shares may result in the company having problems in borrowing money from the banks and in extending its equity capital by selling new shares.

The voice method is easier to exercise when ownership is concentrated so that there are only one or a few major owners. Partly this will give the owners a significant influence and insight (through the number of voting rights they hold at the general meeting and through representation in the enterprise's board), and partly it will generate in them a strong interest in involving themselves in the operation of the company.

If, instead, ownership interests are split among many small shareholders, each shareholder's involvement will be that much smaller. With the absence of a well-defined group that has a great interest in the operation of the company, the demand on the company management to maximize profits will be less urgent. Thus, much would seem to indicate that many small owners will put less pressure on company management than a few relatively large professional owners.

What will this mean for privatization in Central and Eastern Europe? One possibility is that ownership will be organized more or less as follows: the new government of the country could establish investment companies (holding companies) whose only task is to exercise the professional owner's role in several different companies. In each country one could imagine between ten and thirty such investment companies. In turn, shares in the investment companies could be awarded free to the population or sold to them.

SUPPLEMENT 16.1. PRIVATIZATION IN PRACTICE

In *Poland* intense discussions have taken place concerning private ownership. In the spring of 1991 a compromise with the following content was discussed: 80% of shares should be sold or given free to the population; the remaining 20% would belong to workers in the enterprise, perhaps for half the price of what the 80% are sold for. It should also be made possible for 10% (of the 80%) to be sold to foreigners. If foreigners should wish a greater share, special permission will be required. Since the Polish authorities wish to have foreign investments, it is expected that such permission will be virtually a formality.

In another model that has been discussed, 70% of the stocks in a large enterprise would be shared among four groups: the employees (maximum 20%), the general public, financial institutions, and general pension funds. The remaining 30% would be distributed in the form of purchasing rights (vouchers) for all nationals. These vouchers could later be exchanged for shares in various private 'privatization funds' (holding companies or investment companies). To begin with the privatization funds should be managed by experienced foreign financial analysts.

Each privatization fund should in turn be able to utilize its buying rights to purchase shares in enterprises, e.g. in connection with auctions where the authorities simply sell shares in the formerly state-run undertakings. The management of a privatization fund should be given incentives so that it can look after the interests of owners. This may mean a dramatic reorganization of the enterprises. The express objective of the privatization funds must be to maximize the value of the shares owned by the fund (the portfolio of stocks).

Privatization of the first seven large enterprises in Poland was very complicated in purely administrative terms. It resulted in the centrally initiated privatization process being stopped. In some cases the management of state enterprises has itself declared the enterprise bankrupt. Through legal hanky-panky the enterprises have again emerged—like a phoenix from the ashes—with new, private owners (see Supplement 16.2).

In Poland the state enterprises pay an extra tax on wage increases above those recommended by the authorities (see Supplement 18.1). Private enterprises are exempt from such taxes. For this reason employees will often exert pressure to introduce privatization quickly.

In *Hungary* privatization initiated by the central authorities has been slow and bureaucratic. Some people would therefore recommend that enterprises should be encouraged to submit their own privatization plans.

In *Czechoslovakia* small businesses are auctioned off. In the first round only the country's own citizens have been permitted to take part. The authorities stipulate a minimum price for each undertaking. If no domestic buyer can be found at this price, the auction can be carried out again with foreigners also being permitted to make bids.

Ownership of the large Czechoslovakian enterprises is to be spread among the population by means of vouchers. The individual citizen can use his vouchers at auctions. In this way between 40% and 80% of the shares in an enterprise can be spread via vouchers to citizens; the remaining 20% to 60% can be distributed or sold to employees (a small share) and to foreigners. Introduction of something similar to the Polish privatization funds, which could be managed by people who can exercise the ownership role in a professional manner, is also being considered.

Certain enterprises are being earmarked for joint ventures with foreigners, who will pay for their shares with hard currency.

16.5. A Closer Look at Investment Companies

But who is to monitor the monitors, i.e. the investment companies? Those who are dissatisfied with an investment company's profits must themselves be able to sell their shares in this company (the exit method). Should a share of the profits in investment companies be channelled to the development of general pension funds, the man in the street will have a further reason for placing demands on the monitors, in other words for demanding that the investment companies do a good job. After all, his own old age pension will be dependent upon the profits made by the investment companies in years to come.

Should one give the employees in a company shares in their own enterprise, or give them a discount when the shares are to be sold? From an efficiency viewpoint the answer will be no. The reason is that one would then introduce other objectives for the enterprise than maximum profit. If the employees as a whole constitute a large group of shareholders, conservation of jobs and of the existing production structure may well be the result. If there is a reduced demand for the enterprise's products, a necessary reduction of the work-force may be difficult to achieve. The will to introduce new technology that would result in increased productivity and fewer employees might be resisted by shareholders who were also employed in the enterprise. Employment objectives should rather be defined at a national or regional level, and the objectives be attained through macroeconomic measures. Finally, if the aim of maximizing profits should become diluted and vague, the management of the enterprise may have difficulty in borrowing capital for new investments.

'But', someone may say, 'won't job satisfaction and thus efficiency increase if the workers own a share of the enterprise?' This may be right as far as small businesses are concerned, where the roles of manager and owner more or less coincide. In such companies the preconditions for the voice method to function will be met (family businesses perhaps being the best example). However, in larger enterprises the individual will find himself more in an exit situation, with so little influence that it is non-existent in practice.

Experience indicates that enterprises that are owned by the employees have seldom been successful. The Yugoslavian system with enterprises controlled by the workers will illustrate this point. In market economies there are relatively few worker-controlled enterprises. If worker control had been a superior form of company management, then, owing to freedom of business establishment in a market economy, this form would have put alternative forms of organization out of business. However, this has not happened. Joint stock companies with mainly external owners have become the typical organizational form for larger companies.

From a dynamic efficiency viewpoint, it is important that enterprises are reorganized, closed down, or sold when they run at a loss. If this does not happen, labour, buildings, machinery, etc. in these enterprises will be immobilized. The more efficient enterprises will be prevented from using these resources, and total production in society will be less than it could have been. If life is artificially maintained in unprofitable undertakings, other enterprises will be forced to pay more than they otherwise would for *their* production factors. (For example, wages will be forced upward by enterprises that can recruit and keep labour with the help of subsidies.)

Thus, there is a need for a legislative code that will ensure that the enterprise's management will file a bankruptcy petition when the value of liabilities definitively exceeds the value of assets. The state should not get involved through subsidies or cheap loans. If, in an economy, there are no 'hard' restrictions on the enterprises' budgets, over a period of time large parts of the country's business sector will become inefficient.

Experience has shown that bankruptcy legislation and hard budget constraints are necessary prerequisites for economic efficiency. This means that, in its treatment of enterprises and banks, society must draw a sharp dividing-line between the world of politics and the market economy.

The dividing-line between politics and economics also has a deeper political/constitutional significance. A clear distinction between political power and economic ownership will prevent too great an interfusion of politics and economics. If, in the process of privatizing the large enterprises, one establishes a small number of investment companies each of which owns many undertakings, a potential instrument will be created for the exercise both of professional ownership and of political power. For this reason, sceptical voices have warned against too rapid an establishment of large investment companies. Only when one is certain that the old powers that be have been excluded for good from influence in such investment companies should they be established.

In other words, depending upon how deeply anchored democracy and political pluralism are, there may exist a conflict of goals between the demand for economic efficiency and the demand for sturdy barriers against the influence of the old guard.

A more traditional conflict of goals is the one between equitable distribution and economic efficiency.

16.6. Distribution

How should the assets owned by the state in the various countries of Central and Eastern Europe be distributed today?

Small businesses can be sold individually. An ever-recurring question is whether the group of buyers ought in some way to be restricted, for example by excluding former members of the powers that be (since they will have gained their purchasing power illegally), foreigners, or enterprises that have not yet been privatized.

In pre-communist times the person who farmed the land was often its owner. Today's agricultural workers on state-owned farms and collectives are in many cases descendants of former owners of the land. Where the land was formerly divided into many small farms, one might consider returning the land to the former owners or their descendants. They in turn could choose to sell their land to the highest bidder.

Where there is no question of giving the land back to the former owners (e.g. previously German-owned land in Poland), the farmers are often regarded as having gained right of use to the land, or at least right of pre-emption, in the course of the years. This would seem to be a deeply rooted attitude in people throughout the world. 'The land to the tiller!' is a slogan that has sounded not only in Central and Eastern Europe.

The same principle can be applied to those who have rented dwellings. In Central and Eastern Europe in practice rental rights to homes have often been inherited, and individual persons and families in the course of the years have built up a factual right of use. If the free takeover of homes from their owners becomes a rule, this will mean an enormous redistribution of wealth, a redistribution that will set its mark on Central and Eastern Europe for several generations to come. A scheme involving right of pre-emption at a reasonable price would seem to be more appropriate here.

Privatization of the large enterprises raises difficult questions. Do the large enterprises belong to those who work in them? The workers themselves will often maintain that they do. And so will the management (particularly if they were appointed in the old period). On the other hand, it can be claimed that these enterprises belong to the people as a whole.

The fact that a person was previously employed in a large enterprise rather than in a small business surely cannot automatically make him a shareholder when the enterprise is turned into a joint stock company.

The argument concerning risk-spreading also applies to the idea of employees owning shares in the enterprises they work at. In Chapter 13 our neighbour decided that he would rather invest his savings as shares in the shirt factory than in the shoe factory where he worked. If the shoe factory were to go bankrupt he would be doubly affected: both his job and his wealth would vanish. From a risk-spreading viewpoint, therefore, it will be better to have one's wealth in a place other than where one has one's job—not least bearing in mind that there could be rather a lot of bankruptcies in Central and Eastern Europe in the years to come.

In practice, a compromise is being reached on this point. For example, the majority of shares in Poland's large enterprises are to be sold, while a minor portion will be offered to employees at a reduced price. In Czechoslovakia the demand for stocks for employees is insignificant (see Supplement 16.1).

To speed up the process, the state could refrain from taking payment for shares in large enterprises. The shares could be 'parked', either directly in households or in investment companies, which in turn would be owned by the households. What would happen then?

1. Privatization would go more quickly, since one would avoid the need to guess future prices of inputs and outputs in order to form an opinion of the value of the enterprise on this basis.

2. The need for potential buyers to procure credit would also disappear. However, in order to achieve efficiency control through the 'exit method', a market for shares (secondary market for trading in equities) would have to be created.

3. The legal situation could be cleared up by first turning the state enterprises into joint stock companies, where the state owned all the shares, and then privatizing them. Thus, the reform policy would become politically irreversible in a relatively short period. It would then be difficult for the powers that be and employees to seize the enterprise at too low a price. In an unclear legal situation it could in fact happen that the management and employees would be able to acquire a right of ownership that could be politically difficult to change later (see Supplement 16.2). Also, individual people and enterprises might be able unduly to influence the distribution of property in their own favour. 'Wildcat' privatization projects have taken place both in enterprises and in combination with 'spontaneous' land reforms, for example in Poland and Hungary.

4. If privatization went more slowly, stronger incentives for establishing new businesses would be created. Potential entrepreneurs would scarcely

sit with their arms folded for years. They would want to exploit existing profit-making opportunities quickly. Such newly established enterprises would compete with the state enterprises, and possibly put them out of business. For society as a whole, this could lead to a waste of resources. A more efficient use of resources would indicate that state enterprises should quickly be turned into private ones and be given the opportunity to compete freely.

SUPPLEMENT 16.2. EXAMPLES OF 'WILDCAT' PRIVATIZATION: POLAND

Here we shall briefly give a couple of examples of 'wildcat' privatization of enterprises in Poland.

Example 1 The director of a state enterprise starts a private enterprise with the help of the powers that be or the administration in the state enterprise. The most profitable orders that come in are set aside for the new private enterprise. Capital and labour in the state enterprise are hired out to the private one for a symbolic sum, or even provided free of charge.

Example 2 The most profitable part of the state enterprise is liquidated. A new private enterprise is established which is nothing more than the liquidated part of the state enterprise. The new private enterprise hires out its services to the state enterprise, at high prices.

In terms of pure distribution policy, such 'wildcat' privatization is reprehensible. From an efficiency viewpoint, however, things may look different. The persons and groups who take over may be the most efficient at handling both the transition to and the new life in a market economy, in the same way as they were able to adapt and be successful in the old system. Here the conflict between efficiency and distribution really comes to a head!

Source: After Grosfeld (1990)

The discussion of the distribution problems, which we have presented as well as we could here, has been conducted in rather a 'dry' tone. However, we know that few economic questions can create stronger emotions than those that are concerned with equitable distribution. 'Liberté, Égalité, et Fraternité' resounded throughout France two hundred years before the 1989 revolution in Central and Eastern Europe.

Today we know little of what new income distribution will emerge in the old planned economies. It is also uncertain how tolerant most people will be of the emergence of a new class of rich entrepreneurs and businessmen. Tolerance levels will probably be lower, the greater the proportion of 'nouveaux riches' who come from the old ruling class.

Today foreign capital is welcomed by most people. But if it should then prove that economic growth in Central and Eastern Europe has been particularly advantageous for foreign investors, how will people react? Will this be a politically acceptable situation in the long term?

16.7. Stability

'We cannot afford to give the enterprises away!' is a viewpoint that is voiced for example in Hungary. Why not? If the state sold the enterprises instead, with the help of the revenues thereby gained it would be possible to reduce the government deficit. Also, by reducing the stock of money in the private sector, inflation should abate. This may be important, but it does not represent any permanent solution to the problem of inflation. The great budget deficits in the state sector and high inflation rates in the countries of Central and Eastern Europe cannot be solved without the supply of new liquidity also being curtailed permanently. Selling state property once and for all will at best provide temporary respite from budget deficits.

This reasoning points to an important circumstance: privatization is not an appropriate or effective measure for solving the short-term stabilization problems in Central and Eastern Europe. Readjustment to cyclical fluctuations, for the purpose of reducing both unemployment and the rate of inflation, will require quite different, macroeconomic measures (see Chapter 5).

It is important to make it quite clear that, if Central and Eastern Europe are rapidly to establish a new structure in the business sector in relation to the countries' comparative advantages and long-term competitiveness, substantial changes in relative prices will be unavoidable. Such changes will lead to a fall in production and the disappearance of enterprises and jobs in some sectors. This will be necessary so that new industries with sustainable growth will be able to expand. Smaller undertakings, especially within the growing service sector, will attract more labour. In addition, increased house-building and necessary investments in infrastructure will create new jobs.

However, the transition to a market economy will not be a bed of roses. The decisive factor in many cases will be whether changes in relative prices are followed by a sufficiently strong supply response. Will the new opportunities for profitable investments, which will open up in response to changes in relative prices, be grasped by enterprising and skilful businessmen? And will the people of Central and Eastern Europe be able to put up a stronger and sustainable work effort? Only to the extent that these two conditions are met will growth in productivity materialize. Ultimately it is only an increase in production per worker (i.e. increased productivity) which over a period of time can ensure a noticeable improvement in the standard of living for the common man and woman.

16.8. Summary

Privatization is an essential part of the transition to a market economy. During this process, as pointed out, many enterprises may fail or have to reduce their work-forces. It is important to note that a bankruptcy does not necessarily entail the disappearance of all the enterprise's jobs. The purpose of a bankruptcy may be first and foremost to reconstruct the enterprise financially, by having the book equity and debt written down. Write-down of debt will in turn result in those who have lent money to the enterprise, i.e. the state bank, having to take a loss. The book equity in the bank will therefore also have to be written down. For this reason, among others, a reconstruction of the banking system will be a necessary part of the privatization programme.

It can hardly be denied that many large enterprises *will* finally have to be closed down. For this reason, too, it is important that small businesses and service undertakings, which experienced discrimination under communism, are now encouraged to expand through rapid privatization and the establishment of new firms. This will be particularly important if the word 'privatization' in the public mind gradually becomes synonymous with unemployment—something that could halt the privatization process. A certain basic protection of the income of households—by means of a coarse-meshed social welfare safety-net—will be particularly important in such a situation. So will the opportunity for retraining for those who lose their jobs.

17

Elements of Accounting

IN previous chapters we have seen that private ownership of the means of production will lead to efficient utilization of society's resources when owners seek to maximize their profits. Thus it will be necessary to know—in terms of figures—the size of profits. By means of accounting, an attempt is made to arrive at realistic estimates of profits and of other key figures relating to the enterprise's economic development. In this chapter we shall try to provide some insight into accounting.[1] However, let us first look briefly at the calculation of profits in an enterprise.

17.1. Calculating Profits

In Chapter 8 we met an investor, whom we can call Ivan, who invested a million roubles in a new shoe factory. Let us assume he has half a million in equity capital and that he borrows an equal amount from a bank. Ivan becomes the manager of the enterprise and in this capacity he earns an appropriate wage. This wage is used in its entirety to cover personal consumption.

Let us assume that Ivan establishes the factory on 3 January of year 1 (which could be 1991). In this year, and in each of the following nine years, total revenues in connection with sales of shoes exceed total expenditures by 150,000 roubles. The expenditures include wages to employees (including Ivan himself as manager), purchases of raw materials, etc., and interest on the bank loan. Each year Ivan stuffs the cash profit of 150,000 roubles into his mattress at home.

After ten years Ivan has a total of 1,500,000 roubles in his mattress. But now he is weary. He has reached retirement age and wants to sell the enterprise. In January of year 11 he sells the shoe factory along with machinery, equipment, and all sorts of inventory. The highest bid of 400,000 roubles is rather disappointing. Since Ivan now has to pay back the bank loan, this means that he will have to take 100,000 roubles out of the mattress, to add to the 400,000 that the sale of the enterprise raised.

---

[1] This chapter is rather more technical in some respects than other chapters in the book, and can be skipped by readers who wish to proceed to the final chapter without disruption.

In this make-believe example, where we have disregarded inflation and taxes, we can precisely determine the size of the profit deriving from shoe production during this ten-year period. Ivan started out with 500,000 roubles in cash and ended up with 1,400,000 roubles in cash ten years later. The number of roubles can be counted. No doubt, his wealth has increased by 900,000 roubles; i.e. the profit was 900,000 roubles. Profit, then, is the increase in wealth between two points in time.

This principle is important. When the invested capital has increased in value, there is profit. When the values do not consist exclusively of cash, the calculation of values—and thus a numerical expression of the profit for the individual year—will necessarily be rough estimates.

After ten years, Ivan's wealth had increased from 0.5 million roubles to 1.4 million. The average increase in wealth per year—or the average profit—was therefore 90,000 roubles, even though Ivan each year stuffed 150,000 new roubles into his mattress. What happened to the 60,000 roubles that make up the difference between these two figures? The 60,000 roubles are the expression of the loss of value that buildings and machines are subject to as time passes. This loss of value was numerically expressed when Ivan got only 400,000 roubles for a factory that he had paid a million roubles for ten years previously.

The loss of value of the means of production is usually greatest at the start. If Ivan had sold the enterprise after a couple of years, perhaps he would have got 800,000 roubles for it. Of the 800,000 roubles, 500,000 would have to go to paying back the loan. That would leave a net sum of 300,000. In addition, there would be two years' cash profits from the mattress of 150,000 roubles each: 300,000 roubles plus twice 150,000 sums to 600,000 roubles. This would mean an average profit per year not of 90,000 roubles but of 50,000 roubles (600,000 less the 500,000 he invested in the first place, divided by 2 years).

When the profit for the individual year is to be calculated, the value of assets (e.g. buildings, machinery, etc.) at the end of the year must be expressed in numbers. Such estimates are always uncertain. It is only when the whole project is terminated, and the values can again be determined by counting the number of roubles, that profit measurement will be relieved of uncertainty, i.e. will be objective and irrefutable.

The fact that profit measurement for going concerns must be based on rough estimates of value may create problems concerning allocation of resources in society. If the owners are going to borrow money they may be interested in elevating asset values; in this way they can hope for better loan terms, since security on loans will be better.

Confronted with (potential) competitors, on the other hand, owners may often be interested in reporting profits that are smaller than they probably are. Large profits in business may indicate that there is insufficient competition; reports of large profits may therefore attract new investors, resulting in increased competition which may make life less comfortable for those who are already established.

In relation to the authorities, too the owners may be interested in reporting poor profits. For businesses taxes represent expenditure, and since taxes are calculated as a proportion of profits, owners will stand to gain from keeping reported profits as low as possible.

17.2. Financial Accounting and Cost Accounting

In practice, then, the concept of profit is unclear. And the purpose of the estimate of profit may affect its size. In order to avoid the unfortunate effects that purpose-determined stipulations of value may result in, the authorities in market economies have therefore required enterprises to prepare accounts. These accounts are called *financial statements* and must be prepared in accordance with detailed statutory regulations. Financial statements are controlled and approved by *auditors*. No one can become an auditor without advanced training in accounting, tax legislation, etc. Also, auditors must be authorized by the authorities, through appropriate bar exams.

Financial statements consist of

- a *profit and loss account*, which shows how the profit for the period has come about;
- a *balance sheet*, which shows what assets and liabilities the enterprise has at the end of the period, usually the end of the year.

The profit and loss account is also called the *statement of income*, while the balance sheet is also called the *statement of financial position*. We will use the names interchangeably.

The financial statements of joint stock companies shall be available for public inspection, and they shall be prepared at regular intervals, usually once a year. Large enterprises are also required to prepare a financial analysis, which shall explain changes in the enterprise's supply of liquid resources.

These two, sometimes three, reports are published in an *annual report*. Usually reports follow the calendar year. Statements published in March 1991 show the profits for the period 1 January—31 December 1990 and the financial status as of 31 December 1990.

Financial statements provide an overview of the enterprise's financial standing at a high level of aggregation. In day-to-day management of the enterprise, *cost accounts* are useful. Enterprises can freely decide the form of such accounts. In a production enterprise, for example, each department may have its own accounts. The manager of the department prepares regular reports which show how much material has been purchased and consumed, whether consumption has been as expected, how many working hours have been spent on production, etc. With monthly reports, management can quickly intervene if costs are increasing unexpectedly.

Deviations from plans or budgets will give a basis for various types of measures. For example, should sales and therefore earnings in a department drop, this will become clear from the monthly cost accounts. Possible measures in such a case might be:

- transfer some workers from this department to others where sales are booming;
- take extra measures in the form of advertising campaigns, more generous discounts to customers, and pressure on staff in the sales department to make greater efforts.

Financial statements give people and institutions outside the enterprise information about the enterprise's economic situation. For example, a person wondering whether to buy shares in a company listed on the stock exchange will usually spend time acquainting herself with the enterprise's economic situation as it is depicted in the annual report. Correct decision-making will depend on the quality of the information presented in the annual report. The quality requirements for information disclosed in annual reports are dictated by the statutory code, which demands that the information represent a true and fair view of the firm's financial situation.

As mentioned, the financial statement consists of two related sets of information, a profit and loss account and a balance sheet. The relationship between these two accounts is shown in Fig. 17.1. The balance sheet shows the value of the firm's various types of assets: cash and bank

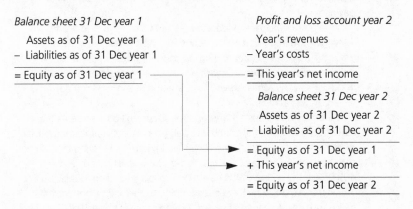

Fig. 17.1. The relationship between the profit and loss account and the balance sheet

deposits, accounts receivable, inventory, production equipment, and fixed properties. Thereafter follows an overview of the enterprise's liabilities: accounts payable, short- and long-term loans from banks and other financial institutions, and direct and indirect taxes payable to the authorities. The value of assets in excess of liability constitutes the firm's equity capital.

The change in equity capital from 1 January to 31 December corresponds to the year's net income. Thus, at any point in time a company's equity capital can be defined as the equity capital paid in at the start of the company plus profits accumulated throughout the years, minus accumulated losses.

If an enterprise runs at a loss, the owners will be the first to lose money. This is expressed by a loss in the profit and loss account and by a reduction in the value of equity capital between 1 January and 31 December. The profit and loss account for the year is also called the 'income statement' (in American English).

SUPPLEMENT 17.1. SOME IMPORTANT TERMS

In the income statement revenues and costs are recorded. Revenues are not the same as in-payments and costs are not the same as disbursements or expenses.

We incur an *expense* when we enter into a binding agreement on the purchase of goods or services. When the expense is paid, we make a disbursement. The time of *disbursement* depends on what terms of payment we receive.

Costs occur when we utilize acquired goods or services in production. They therefore represent consumption of inputs measured in money, while the associated expense is linked to the acquisition itself.

Revenue occurs when we reach a binding agreement on the sale of goods or services. When the bill for the good or service is settled, we receive an *in-payment*.

A special form of costs is *depreciation*. Depreciation represents the amount of money that must be withheld so that the production capacity of assets with a long lifetime will be maintained. In the example of Ivan's shoe factory, the value of the factory fell from 1 million roubles to 400,000 in the course of ten years. If, by paying 1 million roubles in year 11, Ivan could buy a factory corresponding to the one he bought for 1 million roubles in year 1, he would have to have withheld 60,000 roubles per year in order to maintain production capacity. We call this 60,000 *depreciation*. It expresses the loss of value of the production equipment through time. In calculating profits, we then say that annual depreciation constitutes 60,000 roubles.

Note, however, that if the factory had been sold after two years the loss of value would have been 100,000 roubles per year, since Ivan would then have received only 800,000 roubles for the factory and equipment. Depreciation in the enterprise's accounts in that case should have been greater at the start if a realistic figure for profits was to be set. In order not to complicate the picture, in what follows we will stick to an annual depreciation in the accounts of 60,000 roubles.

17.3. Some Principles of Accounting

The profit and loss account shows the financial effects of the year's activities. What activities should be included and how they should be measured is a matter of legal differences across countries. However, the main principles are quite similar. The most important principles are called the principles of prudence, historical cost, and matching.

The *principle of prudence* is based on the philosophy that positive surprises are preferable to negative ones. In practice, this results in expected costs being included in the accounts as soon as we are aware of them; expected revenues, on the other hand, are not included before we are relatively sure that they will be realized. In the balance sheet this principle means that possible increases in the value of liabilities and possible decreases in the value of assets are included as soon as they are expected; possible increases in the value of assets and decreases in the value of liabilities are not included before they are a fact.

The *historical cost principle* means that all valuations must be carried out on the basis of those values that were registered at the time of acquisition. If we purchase raw materials today, the costs in the profit and loss account shall not be corrected even if the price of raw materials has changed by the time we come to use them. The same principle applies to assets that have a long lifetime. Even if the value of a fixed property rises, neither depreciation nor the values in the balance sheet will be adjusted upward. However, unexpected and lasting decreases in the value of assets shall result in a reassessment of the balance sheet values and depreciation sums. This follows from the principle of prudence.

The *matching principle* means that the year's income statement shall report those revenues that have been earned in the course of the year with the deduction of those costs that have been incurred in connection with these earnings. Thus, entry of costs shall be adjusted to those revenues created by consumption of inputs. This will ensure that revenues and costs in the profit and loss account are linked to the same activities. For Ivan's shoe factory, the profit was calculated at 90,000 roubles, after deduction of 60,000 roubles in depreciation per year. Only then had he taken into account the relevant costs.

The general rule is that revenues are counted as earned when binding agreements concerning sale have been concluded, at the same time as the good or service that the contract applies to has been supplied. If Ivan enters into an agreement in December of year 2 concerning the sale of a consignment of shoes, while production is to take place in year 3, he must wait until year 3 before entering the transaction as revenue.

If a good is paid for in advance, the customer's demand to have the good supplied the next year shall appear as a liability in the accounts. However, it is most common for sales to take place on a credit basis. Our demands on customers should then be entered as assets in the balance sheet, in the form of accounts receivable.

17.4. Ongoing Bookkeeping

The bookkeeping system gives an enterprise a complete overview of all its economic activities. It consists of a number of accounts. Each account deals with a particular type of activity. For example, there are separate accounts for wages, leather, rubber, buckles, interest expenses, machinery, cars, sales of finished products, etc.

In the profit and loss accounts we register revenues and costs, while the balance sheets register assets, liabilities, and equity capital. The bookkeeping system must be systematically structured, and all registrations must be accompanied by *documents* such as receipts and invoices. Thus, outside persons such as auditors and tax authorities can check that the accounts are correct, and if necessary can reconstruct the entire accounts from scratch.

Generally, any enterprise can be described in terms of two main flows: one into and one out of the enterprise. These flows are registered by *double entry bookkeeping*, which encompasses the bilateral nature of the flows. In each account we can register both positive and negative flows. Positive flows are entered under 'Debit' in T-accounts or with positive signs in positive/negative sign accounts, whereas negative flows are entered under 'Credit' in T-accounts or with negative signs in positive/negative sign accounts. In T-accounts 'Debit' is on the left, while 'Credit' is on the right. Supplement 17.2 provides some simple examples of how economic activity can be described in a bookkeeping system.

SUPPLEMENT 17.2. EXAMPLES OF BOOKKEEPING

Whereas previously we let Ivan stuff 150,000 roubles each year into his mattress, here we will make the more realistic assumption that this money remains as cash in the enterprise's safe. It would have been better still if he had put the money in the bank and earned interest on it, but to keep the example simple we will let it stay in the safe (at zero per cent interest).

Bookkeeping is done using positive/negative sign accounts, and each account has a special name explaining what the account represents. Activities in year 1 are described under the first eight entries. Entry 9 is made when the account for year 1 is to be closed. All figures are given in thousands.

| Entries | Balance sheet | | | | | Profit and loss accounts | | | | |
	Cash on hand	Factory	Accounts payable	Loans	Equity	Goods pur- chased	Goods sold	Wages	Interest	Dep- reci- ation
1	+ 500				– 500					
2	+ 500			– 500						
3	– 1000	+ 1000								
4			– 350			+ 350				
5	+925						– 925			
6	– 350		+350							
7	– 400							+ 400		
8	– 25								+ 25	
9		– 60								+ 60
Sum	+ 150	+ 940	0	– 500	– 500	+ 350	– 925	+ 400	+ 25	+ 60

Entry 1: The enterprise is established.

Ivan pays in 500. The cash provisionally lying in the enterprise belongs to Ivan. In the accounts this is shown by the fact that the enterprise, which for accounting purposes is treated separately from the owner, owes the owner a corresponding sum. In T-accounts in-payment of cash would be registered under Debit, whereas the owner's claim on the enterprise would be registered under Credit:

Cash on hand	Equity capital
500	500

Entry 2: Loans

Ivan borrows 500 from the bank. This is registered by adding 500 to the cash on hand, at the same time as we register on the loans account that the enterprise owes the bank 500.

Entry 3: Purchase of the factory
This purchase represents an out-flow of cash from the cash on hand and the in-flow of a building worth 1,000.

Entry 4: Purchase of goods on credit
By purchasing goods the enterprise builds up its stock. The supplier is to receive 350 in payment for the goods, but settlement will take place later.

Entry 5: Sales of goods against cash payment
When goods are sold (for 925) there is a physical flow of goods out of the enterprise. In return, the enterprise receives cash from the customer.

Entry 6: Payment for goods purchased
When the enterprise pays for goods purchased the supplier's claims are deleted. The enterprise sends off cash, while the in-flow is the receipt that the enterprise gets from the supplier which shows that the debt has been paid.

Entry 7: Payment of wages
Wage payments involve an out-flow of cash. In return work is performed, and the value of the labour contributed by the workers flows into the enterprise.

Entry 8: Payment of interest
The enterprise must pay 5% per year in interest on the loan. This means that each year the enterprise must pay the bank 25. This is registered as an out-flow from the cash on hand and an in-flow in the profit and loss account for interest. We can imagine the in-flow into the interest account as being the value of the service we buy from the bank in that its money is at our disposal.

Entry 9: Factory depreciation
The year's decrease in value of the factory is estimated at 60, and the value of the factory is reduced by this sum. At the same time, this sum is entered as a cost in the profit and loss account for depreciation. We can regard this as a numerically expressed estimate of the benefit the enterprise has derived from using the factory in the current year. By construction, the sum of the figures in the bottom line of the table is zero.

We have now registered the year's events in the bookkeeping system. With many hundreds of accounts and numerous transactions entered into each account, we would soon lose any overview if the accounts merely consisted of lists of all transactions. In the financial statements, therefore, the information is aggregated and presented in a more user-friendly form.

17.5. Accruals and Deferrals

In the financial statements, the matching principle is employed. This principle is concerned with two concepts: accruals and deferrals. The 'accruals' concept means that revenues and costs are being dealt with in the accounts for the period in which they are earned or incurred. The 'deferrals' concept means that the registration of revenues and costs is postponed to a later date. In practice, the enterprise must make a count of all its assets, rights, and liabilities at the expiry of the accounting period. Thereafter the registrations made in the period must be corrected so that all revenues and costs concerning the accounting year are included. Entry 9 in Supplement 17.2 is an example of a deferral. The decrease in value of the factory is not registered continuously. In order to reach a correct estimate of the year's profit, we must take into account the decrease in value of the factory when the accounts for year 1 are to be finished.

Accruals and deferrals can appear in a number of variants. In addition to depreciation, accruals and deferrals may apply to:

- *Telephone and electricity expenses* These are regularly paid in arrears, after consumption has been measured. However, for the profits for the period to be correct, they must be included as they are incurred.
- *Goods in stock* If the enterprise has not sold all the goods that have been purchased, the profit and loss account for goods purchased must be corrected so that revenues will not be compared with too high an estimate of incurred goods-related costs. This is done by setting up an account in the balance sheet called Goods in Stock.
- *Tax* The enterprise pays income tax on the profit for the year. Payment of tax is a liability that must be included in the accounts in the year that it is demanded.

When the year's revenues and costs are correctly matched, the profit and loss account and balance sheet can be set up. This takes place when the net sums in each account are entered in the profit and loss account or the balance sheet, depending upon the kind of account involved. On the basis of the example in Supplement 17.2, the income statement and the balance sheet have been set up in Fig. 17.2.

The financial statement for year 1 is now finished. The income statement shows how the enterprise's profit in year 1 has come about, and the

Income statement year 1	
Goods sold	925,000
Goods purchased	−350,000
Wages	−400,000
Interest	−25,000
Depreciation	−60,000
Net income	90,000

Balance sheet as of 31 Dec year 1	
Assets	
Cash	150,000
Factory	940,000
TOTAL ASSETS	1,090,000
Liabilities and Equity	
Loans	500,000
Equity	500,000
Net income	90,000
TOTAL LIAB AND EQUITY	1,090,000

Fig. 17.2. Financial statement for Ivan's shoe factory, year 1

balance sheet provides an instant picture of the enterprise's economic standing as of 31 December of year 1.

All types of accounts follow the principle that ingoing inventory plus inflow equals out-flow plus outgoing inventory. For example, the amount of raw materials in stock at the end of the year is counted up and their value determined. When we know how much has been purchased in raw materials in the course of the year and the value of the stock of raw materials both at the start of the year and at the end of the year, the consumption of raw materials throughout the year can be calculated. This method is much simpler to put into practice than having to record the purchase value of every single good that is sold.

17.6. An Example from Real Life

As far as large joint stock companies listed on the stock exchange are concerned, the annual report contains far more information than that required by law. We can see an example of this if we look at the Swedish Volvo Group's annual report for 1990.

The annual report consists of seventy-five pages of information on the Volvo Group including an appendix of seven pages which is aimed at readers unversed in accounting. The annual report itself can be divided into three parts. In the first part the management gives its assessment of the year gone by and the challenges and opportunities that lie ahead.

In the second part we find the economic information on the Volvo Group and on the parent company, AB Volvo[2], including the profit and

---

[2] AB stands for 'Aktiebolag', or joint stock company. The group accounts show the economic standing of the parent company with subsidiaries. A subsidiary is a company where the parent company's share of ownership constitutes more than 50%. The share of

loss account and the balance sheet. Also, figures in the accounts are assessed and analysed. The third part contains information on the employees in the Volvo Group and an evaluation of developments in Volvo shares.

SUPPLEMENT 17.3. VOLVO RAN AT A LOSS IN 1990

Companies are required by law to provide information on developments over a period of time. In a modified form, where we include only the figures for 1990, the Volvo Group's profit and loss account looks like this (all figures in Skr millions):

	1990
Sales revenues	83,185
Costs for goods sold, etc.	– 68,756
Sales and administration costs	– 11,241
Depreciation	– 2,621
Operating profit	567
Reorganization costs	– 2,450
Share of profits in affiliated companies	1,322
Financial revenues	3,965
Financial costs	– 3,731
Profit after financial revenues and costs	– 327
Minority interests' share of year's profit	40
Profit before taxes	– 287
Taxes	– 733
The year's net profit	– 1,020

In the profit and loss account several types of profit are shown. *Operating profit* shows the profit from the group's normal activities, i.e. production and sales of private cars, trucks, aircraft engines, etc.

The item designated *reorganization costs* shows that the group has spent Skr 2.4 billion on improving efficiency and on adapting the group to changed market conditions. The share of profits in affiliated companies shows the group's share of profits in companies where AB Volvo owns between 20% and 50%.

Financial revenues and costs are shown separately. Thus, we can see

the values in the subsidiary that do not belong to the parent company is shown in the item for minority interests. Companies of which AB Volvo owns between 20% and 50% are partially included in the group accounts when ownership is long-term. Such companies are called *affiliated companies*.

what capital revenues the group gains from its financial investments, and what interest costs the loans entail.

The minority interests' share of the profit or loss for the year shows how much of the profits or losses in the subsidiaries have been included earlier in the accounts but do not belong to the group since AB Volvo does not own 100 per cent of all the subsidiaries.

The group has to pay tax, even if it has run at a loss this year. This is because it is the individual companies in the group that are taxed, not the group as a whole. Profits in one subsidiary cannot simply be balanced against losses in another subsidiary. Thus, companies making a profit can be taxed. However, within a company, losses one year can be deducted from taxable profits in later years.

The net profit or loss for the year shows how much the Volvo Group has earned or lost in the course of the year when all relevant types of revenues and costs are included. The group's profit and loss account for 1990 was not satisfactory, since it showed a loss. In the first part of the annual report the reasons for this are thoroughly discussed. Here there is also an account of the measures implemented by the Volvo management to redress the balance.

A modified version of the Volvo Group's balance sheet looks like this (here too only including the figures for 1990):

Assets

Cash on hand and bank deposits	17,585
Accounts receivable	15,718
Goods in stock	19,886
Property, machinery, equipment, etc.	18,327
Long-term ownership shares in other companies	21,530
Various other assets with long lifetimes	9,051
TOTAL ASSETS	102,097

Liabilities and Equity Capital

Accounts payable	7,115
Customer advances	2,122
Various short-term loans	24,521
Short-term bank loans	14,954
Long-term loans	17,794
Minority interests	300
Equity capital	36,311
Net profit/loss for the year	− 1,020
TOTAL LIABILITIES AND EQUITY CAPITAL	102,097

On the assets side, items like cash on hand and bank deposits come first. Items that have a long lifetime of use are recorded last. On the liabilities and equity capital side there is a corresponding order: those loans that must be paid back first come before those loans that run for many years. Finally, there is equity capital, i.e. the difference between total assets and total liabilities.

17.7. Accounts and Budgets as a Basis for Decision-Making

Volvo ran at a loss in 1990, so management has to ask: What can be done better? This question cannot be answered on the basis of the accounts as presented in Supplement 17.3. The aggregation level is too high to know which products sold poorly in 1990 or which departments were run less efficiently. By using cost accounting and budgets, however, management can get a better grasp of the situation.

The budget shows the economic consequences of the activities planned for the future. For example, Volvo might be aiming at selling 76,000 private cars at $US30,000 each in the coming year. With this sales target, the budget can be set up. The budget is usually a profit and loss account designed somewhat differently from the one given above.

Variable costs are costs that vary directly with quantity produced. Included in a car, for example, are one steering wheel, four shock absorbers, two windscreen wipers, two front seats, one brake pedal, etc. Such inputs are called direct materials. With a budget for the number of cars sold, we will also know the expected consumption of such inputs. In addition, labour is needed to assemble the parts into a car. The hourly consumption per car measured in money is called direct wage costs.

Inputs that vary with production but cannot easily be assigned to the individual car are called *indirect costs*. Examples might be consumption of rubber mouldings, emery paper, and simple tools.

Fixed costs are costs that do not vary with production. Should the enterprise halt production for a month, the fixed costs will not be reduced. On the other hand, these variables will disappear when production is terminated.

The distinction between fixed and variable costs depends upon what time-period is considered. In the long run all costs are variable. To see this, think of a salesman who has both a fixed wage per month and a commission calculated on the basis of sales results achieved. The commission will be a variable cost, while the fixed wage will be a fixed cost. However, the salesman can lose his job, and in this perspective the fixed wage will also be a variable.

Let us assume that the sales revenues from exports of Volvo 442 sedan to the United States are estimated at Skr 20 million per month and that each cost component is calculated as shown in Table 17.1. The *contribution margin* (Skr 4 million) shows revenues minus variable costs, i.e. to cover fixed costs (Skr 3 million) and to yield a profit. The expected profit per month here is Skr 1 million.

TABLE 17.1. Monthly sales and production budget (in Skr thousands)

Sales revenues	20,000
Direct material costs	–8,000
Direct wage costs	–6,000
Indirect costs	–2,000
Contribution margin	= 4,000
Fixed costs	–3,000
Profit	= 1,000

Imagine now that the revenues and variable costs for a month were to fall by a total of 40 per cent. The contribution margin would be reduced from Skr 4 million to Skr 2.4 million, and there would be a loss of Skr 600,000.

In such situations the distinction between fixed and variable costs is extremely useful. The fixed costs are unavoidable in the short term. Therefore, in a short-term perspective we should maintain production and sales as long as the contribution margin is positive. A budgeted loss of Skr 600,000 is preferable to a loss of Skr 3 million (which we would suffer by not producing anything at all). When the contribution margin is negative production should come to a halt. With temporary lapses in sales, a clear understanding of the difference between fixed and variable costs is important.

If we now assume instead that sales are in accordance with the budget, but that costs have developed differently than expected, the situation is as depicted in Table 17.2. The number of cars sold corresponds to the planned figure, but the profit is nevertheless Skr 100,000 lower than expected. Material costs are Skr 800,000 higher than budgeted, while

TABLE 17.2. Monthly budget and actual outcome (in Skr thousands)

Sales revenues	20,000	20.000
Direct material costs	–8,000	–8,800
Direct wage costs	–6,000	–5,300
Indirect costs	–2,000	–2.000
Contribution margin	= 4,000	=3,900
Fixed costs	–3,000	–3,000
Profit	= 1,000	= 900

wage costs are Skr 700,000 lower than budgeted. What are the reasons for these deviations?

Closer analyses may reveal, for example, that the wage costs have been reduced because some employees have left their jobs without new workers being engaged. The increase in material costs may be due to increased tempo for the workers, resulting in greater spoilage in assembly. Thus, more parts than normal were rejected. As a consequence, the management should see to it that the employees who left their jobs are replaced, since the savings in wages are more than balanced by increased material costs.

Detailed control of revenues and costs makes it possible for management rapidly to take corrective measures when the deviation between budgets and actual outcomes becomes significant.

Returning to Volvo's annual report for 1990, it is clear that the loss is due among other things to a substantial decrease in sales of private cars. However, we can also draw the conclusion that Volvo's cost accounts have functioned well. At an early stage the Volvo management received signals from the cost accounts that things were not going their way; in autumn 1990 (half a year before the annual report was presented) management decided to spend Skr 2.45 billion on reorganizational measures. Next year's annual report will reveal whether the measures taken to increase efficiency were sufficient to generate profits again.

17.8. Summary

If producers in a market economy are to make good decisions, they have to get their figures right. In this chapter, which is no more than a small foretaste of the subject of accounting, many new concepts have been introduced. In our context—i.e. the transition from the planned to the market economy—it is important that proper accounting systems rapidly come into use. Furthermore, producers, banks, and the authorities should all learn how to utilize accounts in detailed analyses of the development of the individual enterprise and the opportunities facing it.

Financial accounts must be prepared in accordance with the laws that are currently in force or will have to be enacted. It is up to the individual enterprise to develop appropriate cost accounts. With proper routines for the preparation of budgets and accounts, the enterprise will be in a better position to make the right business decisions, targeted at achieving the greatest possible profit.

The workings of the capital market also presuppose that individual enterprises are able to present accounts that give existing and potential investors relevant and reliable information. The case of Volvo clearly demonstrated that, by utilizing its own accounting systems, the enterprise was able at an early stage to recognize the problems surrounding sales of private cars, to implement measures to improve the situation, and to give the general public access to all information via an exhaustive and reliably documented annual report.

18
From Plan to Market: The Challenges

> . . . there is no alternative to shifting to the market.
>
> Mikhail Gorbachev

AT the end of the 1980s, it suddenly became generally accepted that economic planning does not deliver what it promises. Or, as it says on the first page of the report *The Economy of the USSR*, 'there is no example of a successful modern centrally planned economy'.[1]

The challenge lies in introducing a system that almost automatically ensures an efficient allocation of economic resources among alternative uses. A market economy based on private ownership, healthy and vigorous competition, and free price-setting would have these desirable properties.

18.1. Gradual Transition: A Poor Alternative

Before we move on to a detailed discussion of this challenge, among others, a discussion that will be organized around four themes (price reform, macroeconomic stabilization, private ownership, and free trade), we would like to advance the argument that a *gradual* transition would be an inappropriate alternative. Since the various elements in a reform programme are mutually dependent upon one another, a comprehensive change of direction seems to be the appropriate policy. A simple example will illustrate this important point.

Imagine that the authorities in a country changing from a planned to a market economy are worried about how people will react if many sectors of the economy are privatized at the same time as prices are deregulated. Against this background, they decide gradually to introduce privatization and deregulate prices.

For illustrative purposes, let this manifest itself by having the former state-owned shoe factories sold to private persons and the price of shoes

---

[1] This report appeared in December 1990, after five months' work. Those responsible for the report were four international organizations: the International Monetary Fund (IMF), the International Bank for Reconstruction and Development (IBRD), the Organization for Economic Co-operation and Development (OECD), and the recently established European Bank for Reconstruction and Development (EBRD). The serious student is recommended to read the 50-page summary with conclusions.

deregulated. However, as far as shirt production is concerned, there is a desire to gather experience from the 'shoe experiment', perhaps allowing market forces to operate here as well, but at a later date. Thus, centrally planned shirt production will be retained for an unspecified period.

What will happen? With the deregulation of the production and sales of shoes there will immediately be a marked price increase. The queues outside the shoe shops will disappear; and, since queuing-time has been eliminated, money will be all one needs to buy shoes. The scarcity pricing of shoes will become a reality.

With increased shoe prices, enterprising private individuals will see the possibility of substantial profits. Shoe production will be stimulated and competition between manufactures will become stiffer. The authorities will see this as promising—the market is functioning 'as planned', one might say.

To begin with, most people will be satisfied. It is nice to be able to choose between shoes supplied by several competing enterprises, and it is a relief not to have to waste time queuing. The higher prices of shoes will be less popular, even if the increased competition gradually forces prices down to some extent.

After a while consumers will become increasingly annoyed. It may be all very well to have a competitive market for shoes, but the time saved queuing here will be almost cancelled out by longer queues outside the shirt shops. The state undertakings will neither have increased production nor improved the quality of their products. And the prices will be at the same low level as previously. Actually, the *relative* price of shirts has fallen because the price of shoes has risen: hence *longer* queues for shirts. Many a consumer who had actually intended to buy a new shirt will be discouraged by the long queue. Instead, she will go to the shoe shop and come home with yet another pair of shoes.

After a couple of years of pressure by voters, the authorities decide that the time is ripe for liberalization of shirt production, too. Some of the same things that happened when shoes were deregulated will take place now: prices will increase, new producers will set up business, quality will be improved, and the product range will be extended.

When queues vanish outside the shirt shops, people will lose much of their interest in shoes. They now realize that the amount of shoe purchases in the last couple of years has actually been excessive, quite simply because this product was available. Since it is now equally simple to buy shirts, the composition of demand will be more balanced. Shoe manufacturers will experience a marked reduction in demand. Prices will be forced down and many enterprises will have to reduce production. Some will go bankrupt.

From the economic viewpoint of society as a whole, it must now be recognized that the investment in shoe factories was too great. New production facilities for shoes are being utilized to only a fraction of their capacity. Factories built for shoe production will have to be converted, at considerable expense, into factories for shirt production.

How much better the chances would have been for balanced development if shoe and shirt production had been privatized and subjected to competition in the market *simultaneously*!

Given that a rapid and comprehensive change in direction, rather than a slow-moving, gradual transition from a planned to a market economy, is what is needed, we shall consider in more detail the challenges that present themselves. Towards the end of the chapter we shall look at the consequences in the short and the long term entailed by such a change in direction, as well as the problems that such a policy will have to tackle.

Figure 18.1 is a convenient starting-point for a discussion of the various elements in the transition process. Although each country in Central and Eastern Europe has its own very special background as far as history, politics, and economy are concerned, we believe that all of them to a greater or lesser extent face the challenges we are discussing here. The purpose of the following discussion is not to give specific advice. It is more modest: to point to some relationships in principle and submit some generalized viewpoints.

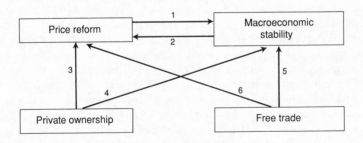

FIG. 18.1. Elements of the transition from plan to market

18.2. Price Reform and Macroeconomic Stability

In the report *The Economy of the USSR* (see fn.1), in the discussion of price reform on pages 24–5, the authors say:

Nothing is more important to the achievement of a successful transition to a market economy than the freeing of prices to guide the allocation of resources. Early and comprehensive price decontrol is essential to ending both the shortages and the macroeconomic imbalances that increasingly afflict the economy. Once resource allocation is based on market clearing prices, the rouble will become an effective medium of exchange; this should largely eliminate the economic disruption resulting from the growth of hoarding, barter, the black market, dollarization and the imposition of restrictions on the free internal movement of goods.

'Price decontrol' simply means that the central planning body ceases to exist, and that free price adjustment based on competition takes over. Then the basis will be laid for prices to function as a signal for allocation of resources, i.e. for *scarcity pricing*, which sees to it that the various resources—capital, labour, raw materials, or energy—are utilized in the most efficient manner.

In Fig. 18.1, arrow 1 reflects the fact that price reform is a necessary prerequisite for macroeconomic stability. By such stability we understand a reasonable balance in public-sector budgets, stable price levels, or at most moderate inflation, and a positive real rate of interest. (See Supplement 12.2 for a refresher on the concept of real rate of interest.)

In 1989 in the Soviet Union price subsidies were of the same dimension as the deficit in the government budget, around 10 per cent of the gross national product (which is the measure of total value added in the economy). Knowing that many prices had remained virtually unchanged since the 1960s—rents had not been regulated since 1928—it is easier to understand the colossal extent of the subsidies. An important element of a price reform is a substantial reduction of subsidies. By greatly reducing the amount of money paid out in this way, a price reform will also contribute to fiscal soundness on the part of the government enterprises.

A price reform, combined with 'hard' budget constraints for large state enterprises, will necessitate better utilization of the resources in such undertakings, and reduce the need for government subsidization.

The transition to free price-setting on the part of undertakings will involve a revolution in the way such enterprises are operated. In Czechoslovakia the state butchers' shops were allowed to set retail prices themselves. Initially, suitable price increases were indicated; after a while it was seen that, at these prices, a number of goods remained on the shelves. Since old habits die hard, a request was made to the authorities for permission to reduce prices. The fact that permission had been granted to do this 'on one's own initiative' had not been realized by those who ran the shops.

Arrow 2 indicates that macroeconomic stabilization improves the possibility of a successful price reform. If market prices are to give participants reliable information, a stable or slightly rising price level will be necessary. A change in the price of a product in such a case will mean that its relative price has changed. Let us imagine that the price of shirts rises by 10 per cent from one year to the next. If the general price level is constant, people will know that shirts have become 10 per cent dearer. If on the other hand the general price level has risen by 15 per cent in the same period, the relative price of shirts will have fallen. However—and this is the point here—when inflation is in the region of 15 per cent it will be harder to maintain an overview of which products are relatively more expensive and which ones have become relatively less expensive.

With moderate inflation, the possibility of maintaining a positive real rate of interest will be improved. When there is considerable competition for investment resources, it will be important for the individual investor to be faced with a price for borrowed funds that reflects the scarcity of capital.

Macroeconomic stabilization involves a balance between public expenditures and tax revenues. When such a balance is established the authorities do not have to resort to the printing-press to finance their expenditures. Then the growth in the money supply can be brought under control, which will make it possible to meet the demand for moderate inflation.

18.3. The Necessity of Private Ownership

Arrow 3, from private ownership to price reform, illustrates Professor Janos Kornai's point taken up in Chapter 4: all experience indicates that co-ordination of economic activity through a free market and free price-setting will be difficult to implement without private ownership of the means of production. When self-interest, which is inherent in private ownership, is given plenty of room, the economy will respond more rapidly to the profit-making prospects that emerge. To put it in another way, supply responses will be more direct, and the economy's adaptability to price changes will be more satisfactory. And Adam Smith's invisible hand will be given better working conditions.

Great challenges must be overcome in order for private ownership to become a reality. First of all, it will be necessary to get a suitable and credible code of legislation into place. In addition to laws that protect—and limit—private ownership, there will also be a need for laws that define procedures in connection with business disputes and bankruptcies.

Second, state-owned enterprises must be transferred to private owners. As discussed in Chapter 16, it may be practical quite simply to auction off the small enterprises. The incentive to run at a profit will then establish itself when private owners take over. And, rather than continuing to subsidize ineffective state-owned enterprises, the authorities will be able to enjoy tax revenues from private enterprises running at a profit.

In addition to transferring ownership to the private sector in this way, such auctions will help to strengthen the government's finances, and to reduce the stock of money in the private sector. A reduction of the so-called *monetary overhang*, or excess liquidity, will contribute to macroeconomic stabilization. This is the reason for arrow 4 in Fig. 18.1.

For larger enterprises—and a characteristic feature of planned economies is that they are dominated by large units—privatization will be a more long-winded and elaborate process. Where it is possible to divide a large undertaking into smaller, more independent, units, it will often be desirable to do so.

Successful privatization of industry will result in more and smaller enterprises. This will provide a basis for stronger competition. In such a situation free price adjustment will ensure the proper allocation of resources. However, there is a limit to everything; if there are *only* small enterprises, economies of scale may remain unexploited.

Another advantage of 'many small' enterprises rather than 'a few large' ones is that the bankruptcy mechanism can be more easily allowed to function; if three or four enterprises, each with 50 employees, have to close down operation, this will cause less of a problem than if one with 500 or 5,000 employees has to do so.

18.4. Free Trade

Referring again to Fig. 18.1, it remains for us to comment upon the box containing the term 'Free trade'. The importance of free trade to macroeconomic stabilization, indicated by arrow 5, rests on the situation regarding competition in the country that previously had a planned economy. Since the process of de-monopolization and privatization will necessarily take a number of years, competition from abroad, i.e. free imports, will be particularly important. With competition from foreign suppliers, domestic consumers will be given alternatives from which to choose. It is the existence of alternatives that is the quintessence of competition. And when domestic alternatives do not exist, foreign alternatives must be allowed to enter the scene, via free trade.

Because of the competition offered by foreign producers, domestic producers will be forced to utilize resources effectively and to maintain competitive prices. Thus, free trade will result in a general downward pressure on prices, i.e. a reduction in inflation. In addition, raw materials and semi-finished products as inputs in domestic production will become more easily available. This will have a stimulating effect on domestic production. And consumers will be able to enjoy a richer selection of products when foreign enterprises are allowed into the domestic market with their goods and services.

The need for free imports is particularly great in sectors characterized by only one or just a few producers. Here the former Soviet Union is an example; Soviet economists have calculated that, out of 5,844 different products, as many as 77 per cent are produced in only one enterprise (see *The Economist*, 13 July 1991: 21).

Arrow 6, from free trade to price reform, indicates that an economy that wishes to take advantage of specialization and international exchange of goods must allow international prices to determine the national prices. In this way the total value of what the country produces will be maximized. Through trade, the country will export goods where it has a comparative advantage in terms of costs of production, and will import goods which would require an inappropriately large consumption of resources in domestic production.

In order for trade with other countries to function properly, a uniform currency rate must be established; that is to say, the price of foreign currency must be the same for all exports and for all imports (see Supplement 15.2). In this connection Poland has organized things properly. The zloty was made convertible with the dollar on 1 January 1990, and the exchange rate was stable until May 1991. Then the zloty was devalued by 17 per cent against the dollar.

Suppose it is desired to reduce consumption of luxury items, such as perfume, video-recorders, and fashionable clothes. Setting quotas on imports of such goods is the worst thing one can do, and customs charges are nearly as bad. If quotas are utilized, those who are lucky enough to have import licences will be in a favourable situation. The scarcity of licences will soon result in the licences having a price of their own; then the owner of a licence will be able to make a gratuitous profit quite simply by selling it. On equity grounds it will be preferable that the authorities sell such licences, for example at auctions, rather than giving them away free of charge. With customs charges, the possibility of such gratuitous profits will disappear, but domestic producers will have less pressure on them to produce goods efficiently.

Since it is the consumption of luxury goods that one wishes to reduce in this case, the correct policy will be to attack the evil at its root, by placing a tax on both domestic and foreign sales of the luxury items in question.

18.5. Incomes Policy in a Transitional Phase?

The leap from a planned economy to a market economy involves enormous challenges, both economic and political. In the first years it will hardly be possible to avoid a considerable number of bankruptcies, a clear rise in unemployment, and a substantial one-time increase in the general level of prices. The fact that some people will have difficulty in finding new jobs is not to be denied. For this reason it will be necessary to establish some form of unemployment benefit quickly. For pensioners and other less privileged groups in society, it will be necessary for a coarse-meshed social welfare safety-net to be in place before the market is 'turned loose'.

However, this bitter medicine will give rise to healthy economic development in the long term. Through bankruptcies, labour will be freed. Enterprises that flourish under the new set of rules dictated by the market will be able to attract new employees. To avoid such enterprises being allowed to let wage levels rise too rapidly, a form of *incomes policy* may be considered.

SUPPLEMENT 18.1. INCOMES POLICY IN PRACTICE

In connection with transition to a market economy, some industries and enterprises may experience very large price increases on their goods and a particularly high profitability. This will provide a basis for significant wage increases. However, if such increases are given, there will be a danger of losing the struggle against inflation; a one-time price increase will be followed in such a case by continuing, high price increases.

Various proposals to avoid such a development have been discussed. Let us briefly consider two of them.

The first one is that the authorities should provide *guidelines* for the wage increases that may be freely allowed. Should an enterprise wish to increase wages beyond this level, an extra tax will be imposed on it. For example, one could say that each rouble of a wage increase beyond the increase stipulated in the guidelines would lead to a tax of one rouble. The cost to the enterprise of exaggerated wage increases would then be twice what the employees received. Such a tax on excessive wage increases would reduce the enterprise's capacity to grant unfettered wage increases, without altogether hindering its freedom to offer workers better conditions.

Another, more drastic, proposal is to set *temporary binding rules* governing permitted wage increases. In this case, for example, the enterprise might be allowed to give workers a maximum of, let us say, 70 per cent compensation for price rises in the previous period. A 'ceiling' on wage increases might possibly be combined with a 'floor', where the enterprise would have to give at least 40 per cent compensation for inflation. Such a system would have a more dampening effect on inflation if wage compensation for price rises were given at six-month rather than three-month intervals.

When the struggle against inflation can be said to have been won, the

incomes policy can be terminated. Instead, wages and other working
conditions can be decided by negotiations between trade unions and
employers' associations.

In Supplement 18.1 a couple of possibilities concerning practical
incomes policy are discussed. Whether a country is well served by such
direct intervention in the setting of wages and prices is a controversial
question. Experience from market-based economies has shown that an
incomes policy may have a *temporary* effect by decelerating wages and
prices; however, after a time (two to five years) market forces will prevail
and wages and prices will accelerate anew. Cases experienced in the
United States in the 1970s exemplify this (see Chapter 9).

The main problem with incomes policies is that prices and wages will
not then reflect the relative scarcity of the various resources. In other
words, an incomes policy will be an obstacle to scarcity pricing. The
result will be a less productive allocation of resources.

Nevertheless, if the economies of Central and Eastern Europe are to
succeed in the fight against inflation, tight regulation of total demand in
the economy will be required. This means that the public sector's total
expenditures must balance its total revenues (fiscal policy), and that the
growth in the money supply must be kept in line with the growth in production (monetary policy). In the short term there is a risk that incomes
policy may become a *substitute for* rather than a *supplement to* tight regulation of the total demand in the economy; if so, incomes policy will not be
conducive to long-term economic growth.

18.6. New Relative Prices

The price rises caused by freeing the market forces will be uneven. Goods
and services that were previously in considerable demand will experience
a far greater rise in prices than the average. Energy, for example, is grossly
underpriced in Russia. The consequence is much wastage of this
resource. When energy prices in Russia move into line with international
levels, this will encourage more economical utilization. It may perhaps be
harder to accept the necessity of higher prices for foodstuffs such as milk,
bread, and meat.

An uneven one-time increase in prices will serve two purposes. The
most important one is to get the relative prices in the economy into order,
so that the scarcity of the various goods and services is reflected in prices.
In other words, scarcity pricing must become a reality. An important consequence of having correct relative prices in the economy is that queues

will disappear. Thus, time will be released for productive activity, and the way will be far better paved for economic growth. The second purpose is to reduce surplus liquidity in the economy. If prices double on average, this means that 400 roubles will now be needed where previously 200 roubles were sufficient. Surplus liquidity means that people in general have a quantity of banknotes in excess of what they need for their daily transactions. It is only when price levels have risen so much that the quantity of notes is 'right' in relation to the economic transactions that take place in society that price levels can be stabilized.

With a healthy monetary system and properly determined relative prices, money will be better able to fulfil its role as a generally accepted means of payment; it will become the 'oil that greases the wheels of the economy', in the words of David Hume.

With competition among alternative suppliers of the various goods and services, and with money as an accepted means of settlement, both firms and households will have less need for holding large stocks of commodities. Since the introduction of the market economy will initially coincide with increased unemployment and falling production, such a reduction in stocks of commodities will help to make it easier to maintain consumption in the economy in the period of transition. Poland, which pulled no punches in its dramatic transition to a market economy on 1 January 1990, experienced precisely that.

Moreover, the decrease in production that Poland registered in 1990 was less dramatic than first impressions might suggest. This is due to two factors: first, what is now produced is more in accordance with what is demanded; second, the growth in production by private enterprises has been registered in the official figures only to a small extent.

When a planned economy which has been closed from trade with Western countries makes way for international exchange of goods, there may be a need for short-term utilization of a common customs charge on imports of finished products to protect the country's own enterprises; some domestic enterprises which in time would be competitive in relation to foreign companies might otherwise risk a demise.

This is connected with the fact that the enterprise must be given time to adjust to the new relative prices, both of inputs in production and of the goods it produces itself. When energy prices in a free market are perhaps trebled in relation to the former 'planned price', it is obvious that the enterprise will receive an incentive to reorganize in the direction of less energy-intensive technology. Further, the enterprise may experience a clear rise in the prices of some of its finished goods while the market does not want others, even at the former price that prevailed under planning. In such a situation protection for a time, where the deadline for zero customs charges has been made known to everyone, may give the enterprise some breathing-space for reorganization, concerning both technology and the composition of inputs in production, and also the composition of the finished goods produced.

How high the customs charge should be, and what sort of deceleration would be most appropriate, are questions with no clear answers. Reasonable economists are at variance with one another on these points. If politicians permit themselves to be influenced to too great a degree by the viewpoints of industry, the result will be too much protection—both in terms of time and in terms of the size of customs charges. In such a case the necessary restructuring of the economy will be retarded.

SUPPLEMENT 18.2. AID FROM THE WEST?

It is perhaps particularly in the areas of technological and capital transfers that one might expect aid from the Western market economies in the reorganization from a planned to a market economy now facing the countries of Central and Eastern Europe.

By 'technological aid' most people would probably understand transfer of modern technology, for example in telecommunications and transport. In many cases technological assistance in rigging up a market economy will be far more important. The establishment of an appropriate code of legislation guaranteeing private ownership, a carefully planned system of taxation when the private sector takes over much of the productive activity in the economy, and a well functioning banking and credit system when investments are no longer decided centrally are enormous challenges that each country must tackle when the market is to replace planning. In these and other areas, the 'old' market economies have valuable experience and technical expertise.

As far as capital transfers are concerned, the previously accumulated debts are an enormous burden to countries such as Hungary and Poland. Relief of this burden of debt will probably be negotiated. For debts to be cancelled, there will be every reason for Western creditors to assert certain conditions on economic policy to be met by the debtor country. Put briefly, these conditions will involve an appropriate transformation in the direction of a market economy. If the necessary reorganization does not materialize, the cancellation of debt could hamper the transition from plan to market rather than provide an incentive to a well planned reform policy. For the country's inhabitants, the harsh measures in the first few years of reorganization might be easier to accept if they were to receive a 'reward' in the form of a reduced burden of debt.

Fresh capital may take the form of gifts, loans, or direct investments. By the latter is meant establishment of foreign firms on a purely commercial basis. Since many new investments may be expected to give good profits in the former planned economies, it is probable that capital transfers will mean quite a lot in the long term. Clearly defined ownership and reasonable political stability are among the prerequisites that must be fulfilled to stimulate direct investments.

Regardless of the time-scale, the most important contribution by the West will be the opening of its markets for imports from Central and Eastern Europe. With free trade, the enterprises of Central and Eastern Europe will adjust to the world market's prices and demands regarding quality, and, in keen competition, will make the best use of their own resources.

18.7. Political Legitimacy

In conclusion, let us consider some problems of a more political nature that will be hard to avoid in the transition from a planned to a market economy.

The transition from plan to market will inevitably involve a redistribution of power. The tasks of the large planning bureaucracy will disappear, and with them the power of those who sit there. The nomenklatura and party bosses will lose their privileges.

It seldom happens that people who are threatened with loss of power and a deterioration in their personal economic situation accept this without resistance. In order to get the former power élite on to the 'market economy team', a professor of sociology in Poland has suggested that these people be given an extra portion of shares when state undertakings are to be reorganized into private joint stock companies. Since the old élite was very adaptable in relation to the former rules, she believes that they will also be able to adapt to the new ones. If this is the case, the most ferocious opponents of the market economy in the past may become trailblazers in the process of reorganization under the new regime.

The problem with such a procedure would be the lack of legitimacy. The fact that those who have profited by the old system should also be given preferential treatment in the new one would naturally be hard for most people to accept.

None the less, with the market in place, it will hardly be possible to prevent some people doing very well economically, while others must stand in unemployment queues for a long time. Large and conspicuous differences in material living standards exist in any market economy. Experience has shown that this may well create a climate of conflict, envy, and dissatisfaction.

If this problem is not to overshadow everything else, it will be important for the new economic system to give relatively quick, tangible, and positive results for a large majority of the public. This is one of the reasons why developments in Poland are being monitored with constant vigilance. The first year of 'the new system' (1990) seems to have given better results than the authorities had initially predicted, despite serious hardships. The queues have vanished. The range of goods is greater. Real wages have fallen by the relatively modest amount of 2–3 per cent compared with three years ago. And economic activity in the private sector grew by more than 50 per cent in 1990 (see *The Economist*, 26 January 1991: 25–7).

In the words of Mikhail Gorbachev, 'The greatest difficulty in our efforts at reconstruction lies in our thinking, which was formed in the past' (1987: 65). Obviously the former Soviet Union is facing greater challenges here than countries such as Poland, Hungary, and Czechoslovakia. While those three countries have lived under a system of economic planning for some forty years, the Soviet Union has done so for a good seventy years. People's attitudes seem to have been affected by this. While most people of Central and Eastern Europe outside the Soviet Union (now the Commonwealth of Independent States) wish to get rid of a system of economic planning that was forced upon them from outside, many of its citizens would seem to be more hesitant about the transition to the new system.

In our view, it is important that the economic challenges resulting from the introduction of a market economy be met not only by politicians, public servants, and experts: there is also a need for widespread dissemination of knowledge of the new system to the people at large. Educators, journalists, and concerned citizens in general have important responsibilities in this regard. With a better understanding of the positive and negative aspects of the market economy, the fear of innovation will recede and a successful transition to a market economy will have a better chance of being accomplished. Knowledge of a strong and rapid river reduces the danger of being stranded in mid-stream.

The final challenge of a political nature that will be mentioned here concerns the legitimacy of the authorities in the eyes of the general public. Such legitimacy is best preserved by democratic elections and political pluralism. Because the introduction of a market economy will require a bitter medicine at first, it is especially important that the authorities enjoy the confidence of, and be accountable to, the general public.

Further, a reasonably clear picture of future changes in economic policy would facilitate decision-making in the public sector. Drawing up, and sticking to, a credible plan for the reform process thus becomes important.

Free, healthy, and vigorous competition is a prerequisite for efficient decision-making not only in the economic sphere, as we have argued in this book, but also in the political arena. Competition is an important source of renewal and progress in political as well as economic life. Just as consumers must have a choice of a variety of products made by competing producers for the economic system to function well, so voters must be free to choose among political programmes offered by competing political parties. In other words, just as the economic system should be geared towards the needs and wants of consumers, as stressed so eloquently by Adam Smith in *The Wealth of Nations*, so the political system should be designed to attend to the interests of voters. Under normal circumstances the discipline of the political market-place, as embodied in a democracy with free elections, will tend to keep politicians, like producers, on their best behaviour for the benefit of the public at large.

References

Abel, I. (1990), 'Subsidy reduction in the Hungarian economy', *European Economy*, 43.

Aganbegyan, A. (1988), *The Challenge: Economics of Perestroika*, Hutchinson, London.

Asselain, J. (1991), 'Convertibility and economic transformation', *European Economy*, special edn. no. 2.

Åslund, A. (1989), *Gorbachev's Struggle for Economic Reform*, Pinter Publishers, London.

——(1990), 'Systemic change in Eastern Europe and East-West trade', EFTA Occasional Paper no. 31.

Begg, D. and others (1990), *Monitoring European Integration: The Impact of Eastern Europe*, Centre for Economic Policy Research, London.

Berlinger, J. S. (1988), *Soviet Industry: From Stalin to Gorbachev*, Edward Elgar, Aldershot, Hants.

Blanchard, O., Dornbusch, R., Krugman, P., Layard, R., and Summers, L. (1991), *Reform in Eastern Europe*, MIT Press, Cambridge, Mass.

Blinder, A. S. (1987), *Hard Heads, Soft Hearts: Tough-minded Economics for a Just Society*, Addison-Wesley, Reading, Mass.

Bofinger, P. (1991), 'Options for the payment and exchange-rate systems in Eastern Europe', *European Economy*, special edn. no. 2.

Burda, M. (1991), 'Labour and product markets in Czechoslovakia and the ex-GDR: a twin study', *European Economy*, special edn. no. 2.

le Carré, John, (1989), *The Russia House*, David Cornwell, London.

Checcini Report (1988), *The European Challenge 1992*, Gower Press, Aldershot, Hants.

Charmeza, W. (1991), 'Alternative paths to macroeconomic stability in Czechoslovakia', *European Economy*, special edn. no. 2.

Cohen, D. (1991), 'The solvency of Eastern Europe', *European Economy*, special edn. no. 2.

Corbett, J. (1990), 'Policy issues in the design of banking and financial systems for industrial finance', *European Economy*, 43.

Crook, C. (1990), 'Perestroika: and now for the hard part', *The Economist*, 28 April.

EBRD, IBRD, IMF, and OECD (1990), *The Economy of the USSR*, Paris.

Erlich, E. (1985), 'The size structure of manufactured establishments and enterprises: an international comparison', *Journal of Comparative Economics*, 9.

Gomulka, S. (1990), 'Reform and budgetary policies in Poland, 1989–1990', *European Economy*, 43.

Gorbachev, M. S. (1987), *Perestroika: New Thinking for Our Country and the World*, Harper & Row, New York.

Gregory, P. R. (1990), *Restructuring the Soviet Economic Bureaucracy*, Cambridge University Press, New York.

Grosfeld, I. (1990), 'Prospects for privatization in Poland', *European Economy*, 43.

Grosfeld, I. and Hare, P. (1991), 'Privatization in Hungary, Poland and Czechoslovakia', *European Economy*, special edn. no. 2.

Hare, P. (1990), 'Reform of enterprise regulations in Hungary: from "tutelage" to market', *European Economy*, 43.

Hayek, F. A. (1944), *The Road to Serfdom*, Routledge, London.

Heilbroner, R. (1990), 'Analysis and vision in the history of modern economic thought', *Journal of Economic Literature*, 3.

Hillmann, A. L. (1990), 'Macroeconomic policy in Hungary and its microeconomic implications', *European Economy*, 43.

Hosking, G. (1990), *The Awakening of the Soviet Union*, Heinemann, London.

Hrncir, M., and Klacek, J. (1991), 'Stabilization policies and currency convertibility in Czechoslovakia', *European Economy*, special edn. no. 2.

Huges, G. (1990), 'Energy policy and the environment in Poland', *European Economy*, 43.

——(1991), 'Foreign exchange, prices and economic activity in Bulgaria', *European Economy*, special edn. no. 2.

Kornai, J. (1990*a*), *Contradictions and Dilemmas: Studies on the Socialist Economy and Society*, MIT Press, Cambridge, Mass.

——(1990*b*), 'The affinity between ownership forms and coordination mechanisms: the common experience of reform in socialist countries', *Journal of Economic Perspectives*, 3.

Landesmann, M. (1991), 'Industrial restructuring and the reorientation of trade in Czechoslovakia', *European Economy*, special edn. no. 2.

Lipton, D., and Sachs, J. (1990*a*), 'Creating a market economy in Eastern Europe: the case of Poland', *Brookings Papers on Economic Activity*, 1.

—— ——(1990*b*), 'Privatization in Eastern Europe: the case of Poland', *Brookings Papers on Economic Activity*, 2.

Montias, J. (1991), 'The Romanian economy: a survey of current problems', *European Economy*, Special edn. no. 2.

Newbery, D. (1990), 'Tax reform, trade liberalization and industrial restructuring in Hungary', *European Economy*, 43.

NOU (1988:21), *Norwegian Economy Undergoing Change*, Universitetsforlaget, Oslo.

Nuti, D. (1990), 'Internal and international aspects of monetary disequilibrium in Poland', *European Economy*, 43.

Phillips, A. W. (1958), 'The relation between unemployment and the change of money wage rates in the United Kingdom, 1861–1957, *Economica*, 25.

Schaffer, M. (1990), 'State-owned enterprises in Poland: taxation, subsidization and competition policies', *European Economy*, 43.

Shiller, R. J., Boycko, M., and Korobov, V. (1990), 'Popular attitudes towards free markets: the Soviet Union and the United States compared', NBER Working Paper, no. 3453.

Shmelev, N., and Popov, V. (1989), *The Turning Point: Revitalizing the Soviet Economy*, Doubleday, New York.

Szalkai, I. (1990), 'The elements of policy for rapidly redressing the Hungarian balance of payments', *European Economy*, 43.

Székely, I. (1990), 'The reform of the Hungarian financial system', *European Economy*, 43.

Uvalic, M. (1991), 'How different is Yugoslavia?' *European Economy*, special edn. no. 2.

Weizman, M. (1984), *The Share Economy: Conquering Stagflation*, Harvard University Press, Cambridge, Mass.

Wiles, P. (ed.) (1988), *The Soviet Economy on the Brink of Reform: Essays in Honour of Alec Nove*, Unwin Hyman, Boston.

Winiecki, J. (1988), *The Distorted World of Soviet-Type Economies*. University of Pittsburgh Press, Pittsburgh.<FRX>

Index